MICHAEL FOGDEN &
PATRICIA FOGDEN

BIRDS OF COSTA RICA

A PHOTOGRAPHIC GUIDE

HELM
LONDON • OXFORD • NEW YORK • NEW DELHI • SYDNEY

This book is dedicated to the memory of its co-author, Michael Fogden, who passed away shortly after submitting it to the publishers. We hope it is a fitting tribute to his lifetime of interest in the birds of Costa Rica.

HELM
Bloomsbury Publishing Plc
50 Bedford Square, London, WC1B 3DP, UK
29 Earlsfort Terrace, Dublin 2, Ireland

BLOOMSBURY, HELM and the Helm logo are trademarks of Bloomsbury Publishing Plc

First published in the United Kingdom 2025

A catalogue record for this book is available from the British Library

Library of Congress Cataloguing-in-Publication data has been applied for

ISBN: PB: 978-1-3994-0663-5; ePub: 978-1-3994-0662-8; ePDF: 978-1-3994-0661-1

2 4 6 8 10 9 7 5 3 1

Designed by Ginny Zeal
Map by Julie Dando
Printed and bound in China by RR Donnelley APS, Dongguan Guandong

FSC www.fsc.org MIX Paper | Supporting responsible forestry FSC® C144853

To find out more about our authors and books visit www.bloomsbury.com and sign up for our newsletters

Cover: Front: Green Honeycreeper (main), Keel-billed Toucan (top left), Rufous-tailed Hummingbird (top centre), Gartered Trogon (top right). Back (top to bottom): Pacific Screech Owl, Boat-billed Heron, Thick-billed Euphonia, Rufous-winged Woodpecker.

CONTENTS

Introduction	4
Map of the region	7
Good birdwatching sites in the region	8
Species accounts	10
Glossary	221
Photo credits	221
Index	222

INTRODUCTION

This pocket guide covers 350 species of birds, representing about 40 per cent of the species that have been recorded in Costa Rica. They have been selected to show those that visitors are most likely to see and the species most typical of the country. The selection includes three of the four Costa Rican endemics (excluding the Cabanis's Ground-sparrow *Melozone cabanisi*) plus a number of Central American endemics. Many North American migrants are common at certain times of year, but those rarely encountered are not included here in order to allow more space for resident birds. The book follows the taxonomy of the IOC World Bird List.

Conservation in Costa Rica

Costa Rica, with the conservation policies that have been put in place, serves as a model for green political action around the world. Costa Rica has more than 30 per cent of its national territory selected for conservation – one of the highest ratios in the world – and has become a leading destination for ecotourism.

Costa Rica is located in between Nicaragua and Panama in Central America. It is bordered by the Pacific Ocean on one coast, and the Caribbean Sea on the other. The regions to the west and east of the continental divide that runs through the country are referred to as the Pacific slope and the Caribbean slope. Costa Rica has a population size of 5.23 million, based on 2023 estimates, and in land area is slightly smaller than the US state of West Virginia, or slightly larger than Denmark.

Costa Rica's climate is defined as tropical and it has mountains rising to 3,821m, including several volcanoes. Roughly 51 per cent of the country is covered with rainforest. While Costa Rica is only roughly 0.03 per cent of the land mass on our planet, it contains five per cent of the world's biodiversity. This makes it one of the most biologically diverse places in the world.

The country's extensive rainforests have become widely known for their remarkable flora and fauna. Attractive animals like Jaguars and other cats, monkeys, sloths, tapirs, macaws, toucans and poison-dart frogs have become particularly celebrated. Costa Rica's diverse habitats are home to more than 830 species of birds, 1,300 species of butterflies and 35,000 species of insects in total. Costa Rica also has roughly 205 mammal species, 220 species of reptiles and 160 species of amphibians.

Costa Rica's geographical position is also important in accounting for its biological diversity. It is connected to both North and South America. As a result, animals and plants have colonised from both continents.

Green policies

Extensive deforestation in Costa Rica between the 1960s and 1980s reduced rainforest cover from 50 per cent to 21 per cent in 1987. But by 2005, the country had boosted its area of rainforest back to 50 per cent. The government has implemented wide-ranging policies to promote the restoration and protection of nature, from eco-hotels and funding for conservation groups to land use policies and formal protection for large parts of the region.

Costa Rica has also funnelled taxes into promoting practices that help protect ecosystems. The government uses taxes from vehicles, fuel and other energy sources to fund management of protected areas and provide environmental services, notably clean air and water. Through a 'Payments for Environmental Services' programme, landowners are paid to protect areas of old-growth forest and to plant new trees. The National Commission for Biodiversity Management, a group comprised of passionate community members, including scientists, politicians and indigenous representatives, acts as an ecological think-tank for the state. This group proposes conservation policies and strategy to the government and disseminates information on sustainability to the wider public.

In 1998, the Costa Rican government implemented the Biodiversity Act, which remains in effect today. This law protects endangered species and thus biodiversity, while giving the state the means to enforce sustainable practices when interacting with biological resources. This means that when land is used for farming, agriculture or industry, protections remain to ensure the health of the natural habitat. Costa Rica has also funded and supported various other groups that promote biodiversity, both publicly and privately.

The role of protected areas

Another aspect of Costa Rica's conservation drive has been to create protected areas. Costa Rica now has 30 national parks, five of which are or comprise part of UNESCO World Heritage Sites, and many other wildlife reserves, sanctuaries and protected wetlands.

Tortuguero National Park, established in 1970, was the country's first national park. Located on the coast of the Caribbean Sea, it includes more than 75,000 hectares of rainforest and waterways, making it one of the largest national parks. It is often referred to as Costa Rica's 'Amazon jungle' because its waterways are the only method of travel. The park is home to plentiful biodiversity with more than 375 species of birds, 400 species of trees and 2,200 other species of plants.

Tortuguero National Park is named after its most important inhabitants – sea turtles. Tortuguero is Spanish for 'region of turtles', and the area was protected as a turtle nesting sanctuary in 1963 before it was declared a national park. The four sea turtle species that lay eggs there use more than 30km of coastline. The park is considered to be one of the best places in Costa Rica for wildlife watching. Despite only being accessible by boat or plane, it is the country's third most-visited park.

Corcovado National Park is the largest national park in Costa Rica and was the second to be founded, in 1975. Located on the Osa Peninsula on the Pacific coast, Corcovado hosts critical biodiversity and is one of the few remaining places where the Baird's Tapir is common. Within it also live a variety of wild cats, including Jaguars, and it is the only place where all four species of Costa Rican monkeys can be seen in the same place. This park has 500 species of trees.

Costa Rica has put emphasis on creating protected areas as a means for conservation. This allows people to safely visit these locations and learn about conservation and biodiversity, and promotes sustainability.

Ecotourism

Costa Rica's focus on living in harmony with nature has benefited the country's economy by boosting one of its largest sectors: tourism. For more than a decade, tourism has represented one of the largest industries in the country, accounting for almost six per cent of GDP in 2015.

Although ecotourism and other forms of sustainable wildlife tourism originated with the environmental movement of the 1970s, ecotourism itself did not become prevalent as a travel concept until the late 1980s. Costa Rica's push towards ecotourism began with the establishment of the Cabo Blanco National Reserve in 1963 and the first national parks in 1970. Ecotourism was seen as a way to protect nature while earning money to finance conservation, as well as to improve living standards through better wages. A 2014 study suggested that the economic benefits of ecotourism contributed to a 16 per cent reduction in poverty in communities living next to protected areas in Costa Rica.

Strong tourism growth began in the mid-1980s and increased significantly in subsequent years, with 800,000 foreigners visiting Costa Rica in 1995, more than one million in 1999, and 2.34 million in 2012. The 2012 visitors generated US$2.4 billion in revenue for the country. Of those visitors, half put their money into ecotourism, including hiking and wildlife watching, as well as visiting rural communities.

Costa Rica promotes and encourages hotels that use energy-conserving practices, notably solar power, and sustainable management practices like water, waste and rubbish reduction. Eco-hotels generally use eco-friendly materials in their buildings, such as locally sourced materials and biodegradable products. Some hotels have volunteer conservation programmes for guests as well as Costa Rican citizens.

As well as promoting sustainable practices, eco-hotels within national parks support the management and maintenance of areas with the parks, while providing employment for local communities.

Earth Law as the next step for Costa Rica?

Recognising that not only is humanity part of the web of life, but also that we need nature to be healthy in order for us to be healthy, is a philosophy that has been put into practice in Costa Rica.

A commitment to creating appropriate legal and tax structures has allowed Costa Rica to show that prioritising nature's health can also benefit the economy, culture and general well-being of citizens. Giving nature a seat at the table, so to speak, by considering the well-being of natural ecosystems along with human welfare, forges a new way forward.

Earth Law could be the next step to ensure permanent legal protections for the gains already made in putting nature first in Costa Rica. Rights of Nature laws would make it a legal responsibility to ensure that nature continues to exist, thrive and evolve. When designated natural areas gain legal recognition, local communities would then be empowered to defend nature against pollution and other destructive forces. Costa Rica could continue to advance its conservation leadership by passing Rights of Nature laws, to continue being a role model for how people can live in harmony with nature.

MAP OF THE REGION

1. Volcán Poás National Park
2. Braulio Carrillo Natioanal Park
3. La Selva Biological Station
4. Tortuguero National Park
5. Cahuita National Park
6. Carara National Park
7. Manuel Antonio National Park
8. Corcovado National Park
9. Monteverde Cloud Forest Reserve
10. Children's Eternal Rainforest
11. Santa Elena Reserve
12. Santa Rosa National Park
13. Guanacaste National Park
14. Palo Verde National Park
15. Lomas de Barbudal Biological Reserve

GOOD BIRDWATCHING SITES IN THE REGION

Costa Rica has more than 60 national parks and reserves, all of which provide excellent sites for watching birds and other wildlife. To find out details of all these sites one should visit the website of the Costa Rica Tourism Institute (ICT) at visitcostarica.com, which provides a wealth of information.

A guide to all the good birdwatching sites in Costa Rica is beyond the scope of this book (for a comprehensive review get a copy of *Where to Watch Birds in Costa Rica*, by Barrett Lawson) but the following areas are particularly recommended. This does not, of course, mean that other sites are not rewarding. The following list simply gives an idea of the many wonderful places that are easily accessible and well worth visiting. If you have only a limited time, visits to La Selva Biological Station, Carara National Park and the Cerro de la Muerte will ensure that you see a good selection of birds in biologically rich habitats.

Volcán Poás is the most visited national park in Costa Rica. The huge active crater is the main attraction but cloud forest rings the beautiful lake in the extinct Botos crater, providing good opportunities to see birds and plants of this habitat.

The **Cerro de la Muerte** has several hotels and is a species-rich highland area with habitats that include oak woodland, cloud forest and páramo. It is a good place to see Resplendent Quetzal, and is close to San José so is easily accessible.

Braulio Carrillo National Park is easily accessible from San José via Route 32. It includes high-altitude rainforest on the Caribbean slope, as well as cloud forest.

La Selva Biological Station is situated on the Caribbean slope in the most biologically rich part of the country. Although the primary function of La Selva is research into the forest and its wildlife, visitors are welcome. As well as the forest, the gardens around the laboratories are good for birdwatching, particularly hummingbirds. La Selva is

Lowland rainforest and inland waterways meet on the Caribbean coast in Tortuguero National Park.

The Río Grande de Tárcoles flowing through Carara National Park.

also an exceptional place to see snakes, frogs and insects.

Tortuguero National Park on the Caribbean coast is a good lowland rainforest site that includes a network of waterways and lagoons. Boat trips offer opportunities for good views of monkeys, sloths, waterbirds and caimans. **Cahuita National Park**, further south, is another good place for birds, with the added attraction of beautiful beaches and coral reefs.

Carara National Park on the Pacific slope, adjoining the south bank of the Río Grande de Tárcoles, is the northernmost outpost of the Pacific rainforest. It is an excellent park for birders where it is easy to see Scarlet Macaws and many of the Pacific forest specialities. The bridge over the Río Grande de Tárcoles is a particularly good place to view the macaws as they fly in to roost in the evening. It is also good for endangered American Crocodiles that doze on sandbanks in the river below.

Manuel Antonio National Park on the Pacific coast is easily reached from San José. It is the most 'touristy' of the national parks but has beautiful beaches, mangroves and easily observed wildlife, including Central American Squirrel Monkeys.

The species-rich **Osa Peninsula** and the protected areas around the Golfo Dulce, notably **Corcovado National Park,** offer a chance of spotting Jaguar and Baird's Tapir, as well as Central American Squirrel Monkeys, Scarlet Macaws, jacamars and turtles.

A trail through Monteverde Cloud Forest Reserve.

The **Monteverde Cloud Forest Reserve, Children's Eternal Rainforest** and **Santa Elena Reserve** are close together and all are easily accessible. The altitudinal range is from about 1,300m in Santa Elena and Monteverde on the Pacific slope up to above 1,600m along the continental divide, and down to about 800m on the Caribbean slope. There are good hummingbird feeders at many of the hotels and the Hummingbird Gallery close to the entrance to the Monteverde Cloud Forest Reserve is also a good place to see Resplendent Quetzal.

The Guanacaste region is a dry forest area that includes **Santa Rosa, Guanacaste** and **Palo Verde National Parks, Lomas de Barbudal Biological Reserve** and good beaches on the Nicoya Peninsula. Santa Rosa National Park is an important wintering habitat for migrant birds from North America, including almost all of the world's population of Scissor-tailed Flycatchers.

Photo key

(♂) Male (Imm) Immature (Nbr) Breeding (Ad) Adult

(♀) Female (Juv) Juvenile (Br) Non-breeding

Great Tinamou *Tinamus major* 43cm

This large tinamou has a short tail, slender neck and rather small head. It is mostly dark olive-brown, faintly barred with black, but often looks plain grey in the field. It is mainly terrestrial, feeding on fallen fruits, seeds, small lizards and frogs, and invertebrates. At night it roosts in small trees. The male alone incubates the lovely blue eggs and cares for the young. The beautiful, stirring song consists of powerful tremulous notes, often given at dusk.

Where to see This is a fairly common resident in lowland rainforest on the Caribbean and south Pacific slopes, locally up to 1,500m.

Highland Tinamou *Nothocercus bonapartei* 38cm

This is another large olive-brown tinamou with a short tail. It is vermiculated with dusky brown and black, and usually looks grey in the field. It feeds on the forest floor, taking fallen fruits and small invertebrates. The call is an un-tinamou-like series of two-syllable nasal notes. As with most tinamous, the male alone takes care of the beautiful blue eggs and young.

Where to see As its name suggests, this tinamou is an uncommon bird of highland forests throughout the country, mostly above 1,200–1,500m.

Little Tinamou *Crypturellus soui* 23cm

This small plump tinamou lacks the barring that is seen on most other tinamous. It is a secretive species that is rarely seen except when walking on trails. Its beautiful voice, however – mellow whistles and trills – is heard frequently. It forages on the forest floor, usually in dense cover, taking seeds, berries, insects, spiders and occasionally small lizards and frogs.

Where to see It is a widespread common resident in rainforest from the lowlands up to 1,500m, throughout the Caribbean and south Pacific slopes, and locally elsewhere.

Black-bellied Whistling-duck *Dendrocygna autumnalis* 51cm

This duck has an upright posture and relatively long legs and neck. It has a russet neck and back, a black belly and a broad white wing-bar, which is conspicuous in flight. The sides of the head are grey, and the bill and legs are pink. Immatures are duller and lack the black belly. It is the largest whistling-duck and the only one with white on the wings. It is gregarious and feeds at night, mostly wading in shallow fresh water, feeding on seeds, young shoots, aquatic insects and snails. Its call is a shrill whistle that can often be heard at night.

Where to see This is a widespread but uncommon resident in the lowlands of both slopes. It is most likely to be encountered in the Tempisque Basin or the Río Frío area.

Blue-winged Teal *Spatula discors* 38cm

The male is distinctive with its white facial crescent, and both sexes have large, pale blue patches on the wings which are conspicuous in flight. Otherwise, the female is rather nondescript, but its small, all-grey bill differentiates it from the larger, massive-billed Northern Shoveler *S. clypeata* (uncommon; not illustrated). Often found in flocks on freshwater ponds and marshes where it forages by dabbling or upending to reach underwater.

Where to see This is the commonest migrant duck, wintering in the Tempisque Basin, the Río Frío area and locally elsewhere in the lowlands and mid-elevations. It is present in Costa Rica from September to April.

Muscovy Duck *Cairina moschata* 86cm

This is a large, heavy-bodied duck. Both sexes are mainly black, glossed with metallic green, except for white wing-coverts. Males are larger than females and have a crest and red caruncles on their face. Immatures differ in having only a small white patch on their wings. It inhabits forested watercourses, including mangroves, and feeds on seeds and grains, including corn and rice, as well as frogs, crab and invertebrates.

Where to see This is a widespread but uncommon resident in the lowlands of both slopes. It is most easily seen in the dry season in the lower Tempisque Basin, particularly in Palo Verde National Park.

Plain Chachalaca *Ortalis vetula 56cm*

A chicken-like bird with a rather small head, a bare red throat when breeding and a long, white-tipped tail. It differs from the Grey-headed Chachalaca in lacking conspicuous rufous primaries. It is found in tropical dry or moist woodland, often in groups of up to a dozen or more birds that are extremely noisy as they fly from tree to tree. The typical call is a three-syllable duet.

Where to see This chachalaca is resident and common in hilly country on the Nicoya Peninsula. It also occurs as a scarce and local resident in the north-west Pacific lowlands as far south as Palo Verde National Park.

Grey-headed Chachalaca *Ortalis cinereiceps* 51cm

The Grey-headed Chachalaca is another chicken-like bird with a long neck and tail. It differs from the Plain Chachalaca most obviously in having rufous primaries which are very conspicuous in flight. It is commonly seen in small flocks of up to a dozen or so birds on the forest edge or in secondary forest, which stay in contact by calling to each other with a loud single shriek or clucking call. Loud duets can also be heard. Whilst foraging, chachalacas often walk gracefully along thin branches.

Where to see This is a resident species and often common in areas where it is not too severely hunted. It occurs in the lowlands and foothills, locally as high as 1,100m, throughout the Caribbean slope and the southern Pacific slope from the Gulf of Nicoya south to Panama.

Crested Guan *Penelope purpurascens* 86cm

The Crested Guan has olive-brown plumage with a prominent bushy crest, pale speckles on the breast and a conspicuous, bare red wattle on its throat. It is a gregarious bird with a loud piping call and is particularly noisy when going to roost at dusk. It feeds on fruit and tender, young foliage in the forest canopy but also descends to the ground to feed on fallen fruits and seeds.

Where to see This guan was formerly resident in the lowlands and foothills on both slopes up to 1,500m or more. Its range has been reduced by habitat loss and hunting. As a result, it is now absent from deforested country and rare in unprotected forest.

Black Guan *Chamaepetes unicolor* 64cm

This glossy-black guan has blue facial skin, a red iris and red legs. During the breeding season, it is noteworthy for its display flight in which it makes a loud rattle with its wings as it glides between trees. It feeds on fruit, including some that are quite large, such as wild avocados and palm fruits.

Where to see Endemic to Costa Rica and Panama, this species is resident in the mountains from about 1,000m up to the upper forest limit. It is common only in remote or protected areas where it is not vulnerable to hunting pressure.

Great Curassow *Crax rubra* 91cm

♀

This is a large, spectacular guan-like species with a yellow cere and knob on its beak, a black-and-white curly crest, and a long tail. The male is glossy black with a white belly, the female mainly rufous. The male's song is a single deep booming note. It is largely terrestrial and often occurs in pairs, foraging for fallen fruits, insects and other small animals.

Where to see Resident in rainforest on both slopes, up to 1,200m in some areas, but becoming scarce and local except in national parks and other protected areas.

Imm ♂

Spot-bellied Bobwhite *Colinus leucopogon* 23cm

This quail is small with a short crest. Both sexes have contrasting dark and white stripes on the head and numerous white spots on the belly that are more prominent on the female than the male, though the female is duller overall. Spot-bellied Bobwhites are terrestrial and live in small groups or coveys of up to a dozen or more birds. They forage in open woodland, scrubby savannah, weedy fields and pastures in search of seeds, berries, insects and grit.

Where to see This species is common in the lowlands and up to 1,500m in the foothills of the dry north-west. Also found in the Valle Central through which their range reaches the Caribbean slope.

Lesser Nighthawk *Chordeiles acutipennis* 22cm

This bird is barred and dotted in browns and greys that make it difficult to spot in its daytime resting position. In flight the dark tips to its wings become apparent and the white band on the primaries (buff on females) is conspicuous. The throat has a white (male) or buff (female) band. Although the bill appears small, the bird has a huge gape that enables it to catch insects easily in flight. It hawks and sallies for moths and other insects from late afternoon through the night until early morning, usually in open, scrubby areas.

Where to see The resident population favours lowland coastal areas on the Pacific slope. The species is also a common autumn migrant, from late September to early November, along the Caribbean coast.

Pauraque *Nyctidromus albicollis* 28cm

This is a common, long-winged, long-tailed nightjar often seen at night on country roads, showing bright ruby-red eyeshine in car headlights. By day, it rests on the ground in woodland where its brown mottled and streaked plumage blends in with its surroundings. At night it hawks for moths, beetles and other insects. Its song is a loud, slightly rough whistle.

Where to see This nightjar is a very common resident, more or less countrywide from the lowlands to 1,500m, or locally even higher.

Dusky Nightjar *Antrostomas saturates* 23cm

This is a dark, rather uncontrastingly patterned nightjar except for its three outer tail feathers, which have white tips on the male, buff on the female. Found in forest edge and clearings in forested mountain regions, where it is widely distributed and locally common. It sallies from dead stumps to hunt for moths and beetles. Its song is low-pitched and only audible at close range.

Where to see It occurs above about 2,000m on most of Costa Rica's mountains but lower, above 1,500m, on the Cordillera de Tilarán.

♂

Common Potoo *Nyctibius griseus* 38cm

This potoo resembles a huge nightjar but, when perched, is easily distinguished by its upright posture. Its plumage is greyish brown, streaked, vermiculated and mottled with blackish and buff. The eyes are yellow, or black with a yellow border depending on pupil dilation; they show brilliant orange eyeshine in the beam of a torch. This potoo is found in forest edge or woodland where it sallies from an exposed perch to catch large flying insects. It calls often on moonlit nights – a series of plaintive whistles, each lower in pitch than the preceding note.

Where to see This species is an uncommon resident countrywide in forested habitats up to about 1,250m.

Great Potoo *Nyctibius grandis* 51cm

This is a large bird, active at night, which appears rather pale in the beam of a torch. Its wings and tail are relatively much longer than those of any owl. It is larger than the Common Potoo with vermiculated plumage (not streaked), and paler, with dark (not bright yellow) eyes and no dark moustachial stripe. The Great Potoo is a bird of lowland rainforest, often attracting attention with its loud calling, a gruff moan that sounds a little like someone vomiting noisily.

Where to see This is a locally uncommon or fairly common resident on the Caribbean slope and in Corcovado National Park on the Osa Peninsula.

White-collared Swift *Streptoprocne zonaris* 22cm

This species is the largest of the swifts occurring in Costa Rica. It is entirely dull-black except for its conspicuous white collar. The tail is slightly notched and often fanned when the bird is soaring. The beak appears tiny but the gape is very large. White-collared Swifts are social, often occurring in large flocks and nesting communally in caves. Like all swifts they are exceptional fliers, enabling them to forage countrywide for their insect food.

Where to see This swift can be seen flying throughout the country. It is mostly seen at low and mid-elevations but tends to roost and nest higher up in the Cordillera Central and Cordillera de Talamanca.

White-necked Jacobin *Florisuga mellivora* 12cm

The male is very smart (and unmistakable) with a blue head, a white collar and a mostly white tail. The female usually lacks these distinctive features and has scaly underparts, similar to the female Scaly-breasted Hummingbird – but with more contrasting white lower belly and vent, and lacking that species' white tail corners. A few females more closely resemble males. Jacobins visit many flowers in the forest canopy, forest edge and treetops in clearing They are also often seen hawking acrobatically for insects above the canopy.

♂

Where to see This hummingbird is fairly common in the lowlands and foothills of the Caribbean slope below 500m and the south Pacific slope below 750m. Seasonal movements make them less obvious in the non-breeding season.

♂

White-tipped Sicklebill *Eutoxeres aquila* 13cm

This is a large, robust hummingbird with streaked underparts and a white-tipped tail. Its most remarkable feature is its very strongly downcurved bill which fits the curved flowers that it visits regularly. It ranges widely in the rainforest understorey visiting appropriate flowers, especially *Heliconia* species with pendant flowers, and *Centropogon*. In spite of its extraordinary bill, it does not have exclusive feeding rights at these flowers. It shares all of them with other hermits, including the relatively short-billed Band-tailed Barbthroat and the Bronzy Hermit. It almost invariably perches to feed, though sometimes hovers at *Centropogon*. Like other hermits, it takes many spiders.

Where to see Found mainly in rainforest in the lowlands and foothills on the Caribbean slope and south Pacific slope, mostly at 300–800m but locally both lower and higher.

Bronzy Hermit *Glaucis aenea* 10.5cm

The Bronzy Hermit is superficially similar to the Band-tailed Barbthroat. Both lack the long central tail feathers of other hermits but boast striking tail patterns – rufous, black and white in this species. The sexes are similar. This hermit's range in Costa Rica is similar to that of the Band-tailed Barbthroat, differing only in being restricted to slightly lower elevations. It is found in forest margins, dense thickets of *Heliconia* and overgrown banana plantations. It is a trapliner and feeds at much the same range of flowers as other hermits. Male Bronzy Hermits are unusual among hermits in not forming leks. The species is also unusual in forming a lasting pair-bond in which males guard the nest, although they play no role in building the nest, incubating the eggs or feeding the young.

Where to see This species is a widespread resident in the wet lowlands below 300m, occasionally 750m. It is easily seen on the Osa Peninsula and at La Selva Biological Station, but occurs at most lowland rainforest sites.

Band-tailed Barbthroat *Threnetes ruckeri* 11cm

The Band-tailed Barbthroat is a typical hermit in its general appearance except for its eye-catching tail, which makes identification easy. The orange gorget on upper breast also differentiates from other hermits. Like its relatives, it is often inquisitive, hovering at arm's length with spread tail, displaying its conspicuous pattern with extensive white base and narrow terminal band. The sexes are similar. It spends more time in the forest understorey than most hummingbirds and is often seen in dense thickets of *Heliconia*. It traplines scattered flowers and is a frequent visitor to shellflowers (*Calathea*), which it pierces, as well as gingers (*Costus*) and *Heliconia*. On the Pacific slope the Band-tailed Barbthroat is one of the few hummingbirds to have a significant song. Alexander Skutch described it as a 'true songster' with a repertoire of plaintive trills and warbles.

Where to see Widely distributed in wet forest on the Caribbean and south Pacific slopes, ascending to about 600m. It is easily seen at many lowland rainforest sites, including Tortuguero, Corcovado and Carara National Parks. The successional plots at La Selva Biological Station are a particularly good location.

Stripe-throated Hermit *Phaethornis striigularis* 9cm

One of the smallest hummingbirds, weighing less than 3g. Its dark eye-mask and downcurved bill differentiate it from all hummingbirds except hermits. It lacks the tail streamers of the Long-tailed Hermit and has a dark tail (mainly white in Band-tailed Barbthroat and red-based in Bronzy Hermit). Though it does not stray far from forest, it readily enters large clearings and gardens. It is a low-reward trapliner, visiting many flowers along a preferred route. Up to 25 males form leks in dense understorey, uttering their squeaky song from low perches, often only 20–30 cm above the ground.

Where to see It is common on the Caribbean and south Pacific slopes and easily seen at all lowland rainforest sites below about 1,500m. Unlike other hermits, it also occurs in evergreen gallery forest in Guanacaste.

Long-tailed Hermit *Phaethornis superciliosus* 15cm

This is a brownish hermit with conspicuous facial stripes, a long, downcurved bill and long, white-tipped central tail feathers. It dwells in the understorey of rainforest, visiting widely scattered flowers along regular routes, often 1km or longer. It will also glean insects and spiders from foliage and spiders' webs.

Where to see Often a very common resident in lowland rainforest and foothills throughout the Caribbean slope and south Pacific slope, sometimes as high as 1,000m.

Green Hermit *Phaethornis guy* 15cm

The male is mostly glossy dark green, tinged with blue on the rump, and its white central tail feathers are short. The female is grey below, with conspicuous facial stripes and a longer white-tipped tail than the male. Much like the Long-tailed Hermit, this species feeds at flowers along a regular route in the forest understorey and gleans spiders and small insects.

Where to see A common resident in cloud forest on both slopes, mainly 800–2,000m, but sometimes lower.

Green-fronted Lancebill *Doryfera ludovicae* 11.5cm

This species is identified by its dark greenish plumage with a white spot behind the eye, shiny bronze crown and nape, and long, slightly uptilted bill. Within its altitudinal range, it is the only hummingbird with a long, straight bill and has more or less exclusive use of canopy flowers with a straight tubular corolla of 30–40mm. It is often seen perched above mountain streams, where it hawks low over the water for insects.

Where to see It frequents wet mid-elevation forest, between 800m and 2,200m on the Caribbean slope and above 1,500m on the Pacific slope. Although it is one of the less common Costa Rican hummingbirds, streams at Monteverde, Tapantí, Savegre and Wilson Botanic Garden are likely places to find it. In Monteverde, the pool below 'La Catarata' on the Río Trail is a good place to look, especially at dawn and in the late afternoon.

Brown Violetear *Colibri delphinae* 11.5cm

The aptly named Brown Violetear has unusually dull plumage for a hummingbird, albeit with blue ear-coverts and gingery rump. Its bill is noticeably short. It is mainly a canopy species, foraging at flowering trees (*Symphonia*, *Inga* and *Calliandra*) and epiphytes (*Clusia* and *Norantea*). It rarely defends a territory but, when it does so, easily holds its own against Rufous-tailed Hummingbirds and other territorialists. During the breeding season, males sing in small leks.

Where to see An uncommon species, it breeds mainly at mid-elevations on the Caribbean slope. It is easiest to see after breeding, when it disperses to lower and higher elevations. From late April to June, small numbers visit the feeders at the Monteverde Hummingbird Gallery.

Lesser Violetear *Colibri cyanotus* 10.5cm

Unusually for hummingbirds, the sexes are alike. They are predominantly iridescent green with a dark band across the tail and distinctive violet ear-patches that are flared in aggressive displays; there are no confusion species at high altitude. The bill is very slightly downcurved. Males can be heard singing repetitively from high in a tree. Like many hummingbirds, the Lesser Violetear is often territorial, but will also trapline many different scattered flowers.

Where to see This species is common in the forested highlands, sometimes as high as 3,000m, but it often prefers more open habitats. A regular visitor to hummingbird feeders.

Purple-crowned Fairy *Heliothryx barroti* 11.5cm

With its streamlined shape and clean-cut colour pattern, the Purple-crowned Fairy is arguably the most elegant and graceful of Costa Rican hummingbirds. The sexes are similar, the male differing in its violet forehead and crown, while the female's tail is longer. It differs from other hummingbirds in its combination of shiny white underparts, white sides to tail, short black bill and contrasting dark eye-mask. It frequents the forest canopy, clearings and flowery gardens, and is usually seen singly. An inveterate thief, it uses its short needle-like bill to pierce and extract nectar through the back of otherwise inaccessible flowers, such as *Erythrina* and *Heliconia*. It even robs the 10cm-long flowers of *Posoqueria latifolia*, a species typically pollinated by long-tongued hawkmoths. More insectivorous than most hummingbirds, it spends much time gleaning from foliage and hawking midges in gaps in the canopy.

Where to see It is widespread on the Caribbean slope up to about 1,200m. On the Pacific slope it occurs as far north as Carara. It is most frequent in the lower mid-elevations on both slopes, between 300m and 800m, but is absent from dry regions, notably Guanacaste.

♂

♀

♀

Green Thorntail *Discosura conversii* ♀7.5cm ♂10cm

This is a small green hummingbird with a conspicuous white band on its rump. The male has an unmistakable long, wiry tail, while the female's is short, but her white moustache and side patches make identification easy. This hummingbird often feeds in the canopy, mainly on small flowers usually pollinated by insects. Feeding birds typically have their tail cocked up at almost a right angle.

♂

Where to see This hummingbird is an uncommon resident, found at 700–1,400m in rainforest on the Caribbean slope. Early in the wet season, from June to August, it typically descends to the lowlands (such as La Selva Biological Station) where it feeds at the flowers of *Warscewiczia*.

♀

♀

Green-breasted Mango *Anthracothorax prevostii* 12cm

The male of this species, with its iridescent green plumage and purple tail, is easily identified. The female, with its broad breast stripe and white-tipped tail, is even more distinctive. Both sexes have a bill that is slightly downcurved. The immature resembles the female but has rufous edging to the white underparts. A few females resemble males. This species spends much time flycatching, often high in the air, but also visits many different flowering trees.

Where to see This is a lowland species that is found in savannah and open woodland, its range extending into deforested areas. It is locally common in northern areas on both the Pacific and Caribbean slopes. Its range is increasing due to deforestation.

Fiery-throated Hummingbird *Panterpe insignis* 11cm

This hummingbird is unmistakable with a good view in good light, though the fiery iridescence on the throat is not at all obvious unless it is seen from head-on and above. Instead, it can be identified by its turquoise rump, blue tail and small white eye-spot. The sexes are alike. Fiery-throated Hummingbirds are unusual in forming a pair-bond of sorts when breeding. The belligerence of males enables them to defend enough flowers to leave a surplus for their partner.

Where to see A regional endemic, this species is common the length of the country in cloud forest areas at high elevations between 1,500m and 2,000m. It is easily seen on Poás and Irazú Volcanoes and the Cerro de la Muerte. There is some movement to lower altitudes after breeding.

Black-crested Coquette *Lophornis helenae* 7cm

The combination of red bill and white band on the rump differentiate this hummingbird from all species except White-crested Coquette; it differs from the latter in its orange-spotted underparts and lack of white on throat or breast. The female has a buffy throat and lacks the male's crest.

♂

Where to see A scarce bird in Costa Rica, found only on the Caribbean slope, mainly between 300m and 1,000m. The most reliable site is possibly the Sarapiquí valley, where it enters gardens to feed at *Stachytarpheta* flowers.

White-crested Coquette *Lophornis adorabilis* 7cm

This tiny hummingbird is sometimes known as the Adorable Coquette. The male is unmistakable with its spiky white crest and dark green cheek tufts that extend back from the side of the neck. It also has a rufous belly and white band across the rump. The female is similar to the female Black-crested Coquette, but has a plain rufous belly and white rather than buffy throat. The White-crested Coquette can be found at all levels of forest and edge, but is most often low, feeding at small flowers, including those of *Inga*, *Vochysia*, *Lonchocarpus* and *Stachytarpheta*. It also takes small insects and spiders.

♂

Where to see This species is quite rare with a restricted range on the south Pacific slope. It occurs from sea level on the Osa Peninsula to occasionally as high as 1,220m. It is endemic to Costa Rica and Panama.

Green-crowned Brilliant *Heliodoxa jacula* 13cm

The spectacular male has a distinctive long, deeply forked tail although this is often held closed so the fork is hard to see. Its plumage resembles shiny plated armour, gleaming from all angles, and its upright stance enhances a rather military look. The female is best identified by its large size, short white stripe below eye, blue tail and green-spangled underparts. Young birds of both sexes, with cinnamon throat and cheeks, cause identification problems for visiting birders and are the reason for many sightings of hummingbirds that are 'not in the book'. Both sexes visit many typical hummingbird flowers, including *Heliconia* and numerous bromeliads. They have strong feet and usually perch to feed.

Where to see Common throughout the wet, forested highlands at altitudes between 800m and 2,000m, it is easily seen at Monteverde, Tapantí and Guayabo. It is abundant at feeders at the Hummingbird Gallery in Monteverde, La Paz Waterfall Garden, and Mirador Cinchona.

♀

♂

Talamanca Hummingbird *Eugenes spectabilis* 13cm

Endemic to the Costa Rica–Chiriquí highlands, this species fills the trapliner niche at high altitudes, where no hermits occur. The male is similar to the Fiery-throated Hummingbird but larger and longer-billed, with a bolder white spot behind the eye and without blue on upperparts or 'fiery' throat. The female is best identified by its size, bill (even longer than the male) and dingy grey underparts.

Where to see It is common in the oak forests of the Cordillera Central and Cordillera de Talamanca at elevations above 2,000m. Also easily seen on Poás and Irazú Volcanoes and along the road that crosses the Cerro de la Muerte. In the latter area, it is a common visitor to the feeders at both Savegre Lodge and La Georgina.

Long-billed Starthroat *Heliomaster longirostris* 11.5cm

This is a fairly large species distinguished from other hummingbirds by its white-tipped tail and long, straight bill. Both sexes have a purplish gloss to their throat, which contrasts with the bold white moustachial stripe. It is a species of the canopy, forest edge, banana and coffee plantations, and open country with scattered trees. It is often seen at the long, blade-like flowers of species of *poró* (*Erythrina*). It also spends much time catching small insects, usually in prolonged darting flights high above the treetops.

Where to see The Long-billed Starthroat is absent from the dry north-west but occupies the rest of the country, from the lowlands up to 1,220m. The closely related Plain-capped Starthroat *H. constantii* (not illustrated) is very similar in appearance. The two replace each other in different parts of Costa Rica, the latter species occupying the dry north-west and Valle Central.

White-bellied Mountaingem *Lampornis hemileucus* 10.5cm

The combination of a white stripe drooping behind the eye, white underparts (with purple throat in male), and bronzy tail with white tips, makes this a distinctive hummingbird. Males are the dominant territorialists within their range, frequenting the canopy, treefall gaps and clearings, where they defend high-quality patches of rubiaceous shrubs such as hotlips (*Cephaelis*). They are virtually certain to be found feeding at any flowering *Quararibea costaricensis*. The smaller females tend to be low-reward trapliners, visiting scattered flowers.

Where to see This hummingbird is endemic to the Costa Rica–Chiriquí highlands. In Costa Rica, it is confined to a rather narrow altitudinal zone in the very wet mid-elevations of the Caribbean slope, from Volcán Arenal southward. On the slopes of the Cordillera de Tilarán, for example, it is common only between 1,000m and 1,300m. It is easiest to see in the Sarapiquí valley, where it is a common visitor to the feeders at Mirador Cinchona (although it is uncommon higher up at La Paz Waterfall Garden).

♀

Purple-throated Mountaingem *Lampornis calolaema* 10.5cm

The Purple-throated Mountaingem differs from the White-bellied Mountaingem in having cinnamon (female) or green (male) underparts. The male's purple throat distinguishes it from the male Grey-tailed Mountaingem, but females are extremely similar. Male Purple-throated Mountaingems are very aggressive. In the lower parts of their range, they are the dominant territorialist, defending the best patches of epiphytic heathers. Though they feed mainly in the canopy, they readily descend into clearings to feed at flowering shrubs, notably hotlips (*Cephaelis*) and *Gonzalagunia*. Females are smaller than the males and forage by traplining scattered or low-quality flowers. Like other mountaingems, the males sing on their territories to attract females. The quiet but attractive song is a medley of splutters, trills and musical warbles.

Where to see This species is common in cloud forest in the northern Cordilleras and the Cordillera Central, most abundantly between 1,500m and 2,500m. It is absent from most of the Talamancas, where it is replaced by the Grey-tailed Mountaingem. Good sites are the Monteverde Cloud Forest Reserve and Braulio Carrillo and Tapantí National Parks. It is a frequent visitor to the feeders at the Monteverde Hummingbird Gallery and at La Paz Waterfall Garden.

♂

♀

♂

Grey-tailed Mountaingem *Lampornis cinereicauda* 10.5cm

This species differs only in tail colour from the closely related White-tailed Mountaingem *L. castaneoventris* (not illustrated). In the past, both were lumped with the Purple-throated Mountaingem as one species, the Variable Mountaingem. The male Grey-tailed's white throat distinguishes it from both Purple-throated and White-bellied Mountaingems; it also has a grey tail, rather than blue and bronzy, respectively. The females of both Grey-tailed and White-tailed are almost identical to the female Purple-throated. Like the Purple-throated Mountaingem, the male is the dominant hummingbird in the lower parts of its range.

♀

Where to see It inhabits oak forest from about 1,800m up to the timberline. Virtually endemic to the Cordillera de Talamanca, it is replaced at the Panama border by the White-throated Mountaingem. It is fairly common on the Cerro de la Muerte and is easily seen at Savegre Lodge, where it is a regular visitor to the feeders.

♂

Magenta-throated Woodstar *Philodice bryantae* 7.5cm

This attractive hummingbird frequents semi-open areas and forest margins. When hovering, it has a characteristic appearance, with a cocked tail, like coquettes and thorntails. Its flight is smooth and beelike, accompanied by the loudest hum of any Costa Rican hummingbird. Males have an impressive diving display flight, each dive accompanied by a snipe-like bleating noise and wing snaps. In addition to cocked tail and noisy flight, large white patches on rump sides and partial white collar also help identify this species. Both sexes are aggressive and territorial amongst themselves, but often forage as filchers, visiting flowers in territories of other hummingbirds.

♂

Where to see It occurs on the Pacific slope, between 1,000m and 1,600m, occasionally reaching the Caribbean slope at passes in the north. Though generally rare, it is locally common in Monteverde, where it is present from October until about May. It is endemic to Costa Rica and Panama.

♂

Ruby-throated Hummingbird *Archilochus colubris* 8cm

The male's red throat and black chin are distinctive. It has a darker face than the male Volcano Hummingbird, bolder white breast band than the male Scintillant Hummingbird, and lacks the Magenta-throated Woodstar's white rump sides. The Ruby-throated's short, blackish tail with shallow fork also differentiates it. The female has whitish underparts and more eye-catching whitish tail-tips than the females of those species. The Ruby-throated Hummingbird is the only Costa Rican hummingbird to make a very long migration, travelling up to 4,000km from eastern North America (including Canada) to Central America. Many individuals fly non-stop for 800km across the Gulf of Mexico, while others go around. It is a notable achievement for so small a bird. On average they weigh about 3.5g, but prior to departing for migration, adults put on nearly an extra gram of body fat to fuel their long journey.

Where to see This species is a locally common winter visitor to the north Pacific lowlands and foothills (including Monteverde), arriving from mid-October to November and departing by late March or mid-April. It is also a rare or uncommon visitor to the Los Chiles/Río Frío and Valle Central areas.

Volcano Hummingbird *Selasphorus flammula* 7.5cm

Male Volcano Hummingbirds on different mountains have different coloured gorgets – rose-red on Volcán Poás and Volcán Barva, purple on Volcán Irazú, and a rather dull purplish grey in the Talamancas. The male is similar to the male Scintillant Hummingbird, but has a mainly black (rather than mainly orange) tail, and its gorget is never flaming red like the Scintillant. The female of all forms resembles the female Scintillant Hummingbird, but is whiter below, less buffy, with more conspicuous white tips to the outer tail feathers. Both sexes are low-reward trapliners and filchers that visit many different flowers. Breeding males defend territories that are often on the summit of small hills. The male has a spectacular display in which it rises vertically above its territory, before hurtling back down, calling and snapping its wings.

Where to see A regional endemic, this species is common on the Cordilleras Central and de Talamanca, from about 2,000m upwards, in particular on Volcán Poás, Volcán Irazú, and the Cerro de la Muerte. It is a common visitor to the feeders at La Georgina.

Scintillant Hummingbird *Selasphorus scintilla* 6.5cm

The Scintillant Hummingbird is Costa Rica's smallest bird, weighing only 2g. Its tiny size rules out almost all species. The male's orange-red gorget and mainly orange tail distinguish it from the male Volcano Hummingbird. The female is very similar to the female Volcano Hummingbird, but it has orange (not green) central tail feathers and the tips of the outer tail feathers are orange rather than whitish. It frequents brushy areas, overgrown pastures and hedgerows, where it feeds mainly at insect-pollinated flowers. Breeding males greet females and intruders with diving displays. In lieu of a song, their outer flight feathers make a characteristic metallic whistling sound when they fly or dive.

Where to see It breeds at mid- to high elevations on the Pacific slope from the Cordillera Central south to Chiriquí in Panama. After breeding, Scintillant Hummingbirds disperse, moving to higher elevations on the central volcanoes as well as north-west along the Cordillera de Tilarán. Small numbers reach Monteverde by the beginning of March, sometimes earlier, and most depart in June. It is endemic to Costa Rica and Panama.

♀

♀

♂

Canivet's Emerald *Cynanthus canivetii* 8cm

There are several closely related populations of small, iridescent green emeralds, ranging from Mexico to northern South America. In Costa Rica, birds from the north Pacific slope are called Canivet's Emerald. The males have a red-and-black bill and a forked tail, which they will pump up and down while hovering. They frequent forest edge, cleared scrubby areas and gardens. Similar birds from south-western Costa Rica are called Garden Emerald *Chlorostilbon assimilis* (not illustrated) – they have a fully black bill and a less forked tail. Females of both species have white underparts and eye-stripe, and a black cheek, but there is no range overlap.

Where to see Common on the Pacific slope below about 1,500m. The Canivet's Emerald is easy to see in the Monteverde community, while the Garden Emerald is frequently seen at Wilson Botanic Garden.

♂

Violet-headed Hummingbird *Klais guimeti* 7.5cm

The male Violet-headed Hummingbird is easily identified, although its violet head can be hard to see. The rather nondescript female is less distinctive. The conspicuous white spot behind the eye, short, straight bill and small size are good identification features for both sexes. This species frequents the forest canopy, second growth and clearings, and readily enters gardens close to forest. Though small and low in the hummingbird pecking-order, males set up territories if they can find flowers that are undefended. They defend them fiercely against other small hummingbirds, as well as bees, wasps and butterflies. Females often forage as low-reward trapliners, stealing nectar in the territories of other hummingbirds.

Where to see It is common below 1,000m on the Caribbean and south Pacific slopes. Any patch of *Stachytarpheta* in the vicinity of forest is very likely to attract a few.

♂

Violet Sabrewing *Campylopterus hemileucurus* 15cm

This is a large, robust hummingbird with a longish, curved bill. Both sexes have a substantial tail with conspicuous white outer feathers. Otherwise, the male is mainly violet and the female dull green and grey with a dull violet gorget. The Violet Sabrewing is similar to hermits in being a trapliner, keeping fairly low in the forest and specialising on scattered flowers with plentiful nectar.

Where to see It is a common resident in montane forest on both slopes between 1,500m and 2,200m, with many birds moving to lower elevations after breeding.

Bronze-tailed Plumeleteer *Chalybura urochrysia* 11cm

This species goes by the name of Red-footed Plumeleteer in some books and checklists, perhaps a more appropriate name for Costa Rican birds since the red feet are unique amongst hummingbirds in a Costa Rican context. The male is mostly green with a bronzy rump, while the female is whitish beneath, and her dark bronzy tail has pale tips. The Bronze-tailed Plumeleteer frequents forest edge, second growth and gaps but seldom strays far from good forest. Males are territorialists, controlling prime patches of such flowers as *Hamelia patens*, hotlips (*Cephaelis*) and species of *Heliconia* with short flowers. Females seldom defend territories but, being large and aggressive, often forage in the territories of other hummingbirds.

Where to see It is widespread in the lowlands of the Caribbean slope, ascending to 700m in the foothills. It seems most numerous in the Sarapiquí lowlands, especially at La Selva Biological Station, where it is common. It is rather shy, which can give a false impression of its status. When its voice is known, a truer picture emerges.

Crowned Woodnymph *Thalurania colombica* 9–10cm

The male's glittering violet and green plumage and deeply forked, blue-black tail make it one of the most beautiful hummingbirds. The female is mostly green with a well-defined pale grey throat and chest and a slight tail fork. This species is found in wet rainforest where in feeds on the nectar of many different flowers. During the breeding season (February to June), the male forages mainly in the canopy while the female concentrates on the forest understorey. After the breeding season, both sexes are very dependent on *Heliconia* flowers.

Where to see This hummingbird is a common resident from sea level to 750m in rainforest on the Caribbean and south Pacific slopes. After breeding it often moves higher, up to at least 1,500m.

Snowcap *Microchera albocoronata* 6.5cm

The Snowcap is tiny. It weighs only 2.5g and is smaller than the coquettes. With its shining white cap and unique maroon-purple plumage, the exquisite male is unlike any other hummingbird. The female, on the other hand, has bronzy-green upperparts and white underparts, and resembles several other small female hummingbirds. It is best distinguished by a combination of tiny size, short bill, and tail with bronzy central feathers and extensive white on sides. It is a canopy species, feeding at trees such as *Inga* and *Warszewiczia coccinea* but descending to shrub level at edges and in clearings.

Where to see From January to May it breeds at lower mid-elevations between 300m and 700m, the length of the Caribbean slope. After breeding, most descend to the foothills. It is not an easy hummingbird to see except at Rancho Naturalista, near Turrialba, where it visits hummingbird feeders. It is quite common in the lower parts of Braulio Carrillo National Park and occurs regularly at La Selva Biological Station in May and June, where it is almost invariably seen at *Warszewiczia*, *Hamelia* or *Stachytarpheta*.

Coppery-headed Emerald *Microchera cupreiceps* 7.5cm

This species is one of the smaller hummingbirds, with a slightly downcurved bill. The male has a copper crown, rump and centre to its tail, while its outertail feathers are conspicuously white. The female has white underparts. It forages at most levels of the forest and ventures out into forest edge and clearings. It is a common visitor to feeders.

Where to see Endemic to Costa Rica, it is a common resident on the Caribbean slope at altitudes of 700–1,500m, lower outside the breeding season. Some reach the Pacific slope via passes in the northern cordilleras. It is, for example, common in the Monteverde Cloud Forest Reserve.

Stripe-tailed Hummingbird *Eupherusa eximia* 9.5cm

The male has a brilliant green throat and breast, white on the lower belly, a rufous patch on the wing and white in its outer two tail feathers (visible in flight but not at rest). The female has greyish underparts, and its rufous wing patch is duller but still conspicuous. The upperparts of both sexes are green, tinged with bronze. They are birds of the forest canopy though they will descend into more open areas in search of flowers. They are generalist feeders, sometimes territorial, sometimes stealing nectar by piercing flowers, and regularly visit feeders.

Where to see It inhabits mid-elevations on both slopes, though it is patchily distributed. It may move to higher or lower altitudes outside the breeding season.

Black-bellied Hummingbird *Eupherusa nigriventris* 8cm

The male has a black face and underparts, while most of the tail is white with darker central tail feathers, and there is a rufous patch on the wing. The female differs in having grey underparts, and is quite similar to the female Stripe-tailed, but has more white in the tail and the rufous patch on the wing is inconspicuous. This species inhabits wet forest, the male favouring the canopy while the female often forages at lower levels at edges and gaps.

Where to see This species frequents the Caribbean slope at low to mid-elevations, between 900m and 2,000m during the breeding season (October to March, though sometimes earlier) but sometimes lower when not breeding. It visits feeders in the Sarapiquí valley. It is endemic to Costa Rica and Panama.

Scaly-breasted Hummingbird *Phaeochroa cuvierii* 12cm

This hummingbird is rather similar in size, structure and behaviour to a sabrewing, and it is sometimes now included in the genus *Campylopterus*, instead of the monotypic *Phaeochroa*. The sexes are alike and rather nondescript, but their large size and the white corners to their tail make them easy to identify. The species occurs on the forest edge, in semi-open areas and in mangroves in Guanacaste. Though sometimes territorial, it often marauds into the territories of others. Favourite flowers include *Inga*, *Erythrina*, several species of *Heliconia*, *Pelliciera* when in mangroves and the terrestrial bromeliad *Bromelia pinguin* in Guanacaste.

Where to see This species is locally common on the Pacific slope, up to about 1,200m in the south-west. It also occurs in the Río Frío area south of Lake Nicaragua, but is rare everywhere else on the Caribbean slope. It is usually easy to find at Wilson Botanic Garden, where there is an active lek for much of the year.

Blue-vented Hummingbird *Saucerottia hoffmanni* 9cm

The isolated northern population of this hummingbird, which is found in Nicaragua and north-western Costa Rica, has been split from South American birds (known as Steely-vented Hummingbird *Saucerottia saucerottei*). Male and female Blue-vented Hummingbirds are similar, although the female is duller. The combination of blue undertail-coverts and tail, and white thigh feathers, is distinctive. Both sexes defend territories, and are often the dominant territorialist at mid-elevations, but subordinate to the slightly larger Cinnamon Hummingbird where they overlap in lowland Guanacaste. They frequent scrubby woodland, coffee plantations and gardens, foraging at flowering trees such as *Inga* and *Tabebuia*, as well as at shrubs and herbs.

Where to see Easily seen throughout north-western Costa Rica, it is found on the Pacific slope north of Carara National Park up to about 1,500m. In the Monteverde area, some birds cross the continental divide and move to mid-elevations on the Caribbean slope around May and June, when *Inga oerstediana* is flowering. A few individuals may be seen at the same time of year in the Sarapiquí valley on the Caribbean slope of Volcán Poás.

Snowy-bellied Hummingbird *Saucerottia edward* 9cm

The male and female of this species are alike and resemble the male Mangrove Hummingbird in having a shiny green breast that contrasts sharply with its white belly: this feature is otherwise unique among Costa Rican hummingbirds. There is little chance of confusion, however, because the Snowy-bellied is a foothill species, found mainly between 300m and 1,600m, and does not occur close to mangroves. It inhabits open woodland, coffee plantations and gardens, where it forages in the canopy of flowering trees and in shrubs, including *Inga* and *Calliandra*. Unlike many of the other hummingbirds alongside which it lives, the male Snowy-bellied Hummingbird sings solitarily and inconspicuously.

Where to see A regional endemic, it has a limited range in south-western Costa Rica, excluding the wetter areas around the Golfo Dulce. Two of the best places to see this species are the El General valley and Wilson Botanic Garden.

Cinnamon Hummingbird *Amazilia rutila* 9.5cm

With its distinctive cinnamon underparts and tail, this hummingbird is easily identified. The sexes are similar, apart from the male having more red on its bill, and both are aggressive territorialists. The Cinnamon Hummingbird is found in dry forested areas, and spends most of the dry season foraging in flowering trees whose main pollinators are bees and other insects. It is attracted to legumes and to the synchronized 'big bang' flower displays of species of *Tabebuia*. In the wet season the big terrestrial bromeliad *Bromelia pinguin* is important when it flowers around April or May.

Where to see It is resident and common up to 500m in Guanacaste, in deciduous woodland, scrub, parks and most hotel gardens. It ranges as far south as the Río Tárcoles and up into the foothills of the northern cordilleras.

Rufous-tailed Hummingbird *Amazilia tzacatl* 10cm

As its name suggests, this species has a rufous tail, which is slightly forked. The rest of its plumage is mainly bronzy-green. The sexes are similar though the female has a less extensive glittering-green breast and less red on its bill than the male. The tail colour is shared only with the Cinnamon Hummingbird, from which it differs in largely green (rather than cinnamon) underparts. It is common in large clearings with scattered trees and open areas. Males commonly defend territories at clumps of *Hamelia*, *Heliconia* and other popular flowers.

Where to see This hummingbird is scarce in the dry forested north-west but otherwise common on both the Pacific and Caribbean slopes, sometimes as high as 1,800m, wherever forest cover has been removed. It also occurs in built-up areas, including gardens in the centre of San José. It is often the first hummingbird seen on any visit to Costa Rica.

♀

♂

Mangrove Hummingbird *Amazilia boucardi* 9.5cm

The male of this species has a bright blue-green throat and chest and a conspicuous white belly. The female is mainly white below, speckled with pale green on its flanks. Other hummingbirds are similar but this species is largely restricted to mangroves where it is unlikely to be confused. Its principal food source is the nectar of Pacific mangrove flowers, though it sometime moves inland to visit flowering trees of *Tabebuia* and *Inga*.

Where to see Locally common in mangroves along with Pacific coast between the Golf of Nicoya and Golfo Dulce, this species is nevertheless classed as Endangered because it has a small, fragmented range and its specialised habitat is itself threatened. It is a Costa Rican endemic.

Blue-chested Hummingbird *Polyerata amabilis* 8.5cm

The male's blue throat and female's speckled blue throat differ from all other Costa Rican species except the Charming Hummingbird (with which it has no range overlap – Charming replaces it on the south Pacific slope) and the Blue-throated Sapphire (which has a mainly red bill and golden tail). It is a bird of second growth, old clearings and river banks, as well as overgrown plantations and shady gardens. Preferred flowers include those of *Inga* and *Warszewiczia*, and rubiaceous shrubs such as *Hamelia* and *Palicourea*. Males gather in small leks and sing with great persistence for much of the year.

Where to see Confined to the humid lowlands and foothills of the Caribbean slope, rarely above 300m, it is easily found in suitable habitat, including at Tortuguero and Cahuita National Parks. At La Selva Biological Station it can almost always be found at dawn at clumps of *Hamelia patens* growing around the laboratories.

Charming Hummingbird *Polyerata decora* 9cm

In some books and checklists this species is known as the Beryl-crowned Hummingbird. It is closely related and very similar in appearance to the Blue-chested Hummingbird of the Caribbean slope, the two species replacing each other geographically on opposite sides of the Talamancas. They are often considered to be the same species. Like most of its relatives, the Charming Hummingbird occurs in open woodland, treefall gaps, coffee plantations and on the forest edge. Both sexes are attracted by a wide range of flowers, including *Inga*, rubiaceous shrubs (*Cephaelis, Hamelia* and *Palicourea*) and *Heliconia*. In the higher parts of its range, above 1,000m, it also feeds at epiphytic heaths.

Where to see It has a very small range, being confined to the south Pacific slope and adjacent Panama, up to altitudes of about 1,200m. Within this range it is common in most suitable habitat and is easily found along the Pacific coast from Carara southward, in the El General valley, on the Osa Peninsula and at Wilson Botanic Garden.

Blue-throated Sapphire *Hylocharis eliciae* 9cm

This species was formerly called the Blue-throated Goldentail. With its lustrous tail, glittering blue gorget and largely red bill, the male is spectacular and distinctive. The female is similar but duller, the blue gorget being reduced to blue spotting. The red on the female's bill is also less extensive. It is a bird of light secondary growth, overgrown clearings and forest margins. Both sexes trapline flowers and rarely defend territories. For eight or nine months of the year, males gather in leks of 3–10 birds, perched on bare twigs, 8–10m above the ground. Their loud, cheerful song draws attention to the leks.

Where to see It occurs over much of Costa Rica, up to about 300m. It is relatively scarce on the Caribbean slope but often abundant in the south Pacific lowlands and foothills, including on the Osa Peninsula.

Smooth-billed Ani *Crotophaga ani* 35cm

This member of the cuckoo family is very similar to the Groove-billed Ani but is a little larger. It is all black, with a long tail and is often rather dishevelled looking. It has a heavy, laterally flattened bill, and the high crown of the upper mandible gives the bill a characteristic humped outline. It feeds mainly on insects, usually in association with livestock and occasionally with army ants. It can often be seen sunbathing with wings spread, as in this photo.

Where to see It prefers open habitats and occurs commonly in the southern Pacific regions of Costa Rica, from sea level up to 1,200m, where it has largely displaced the Groove-billed Ani.

Groove-billed Ani *Crotophaga sulcirostris* 30cm

This species is an all-black member of the cuckoo family with a heavy, grooved bill that is compressed laterally and gives the bird a characteristic appearance. It has a long tail, which is floppy and appears loosely attached. It is very similar to the Smooth-billed Ani but has a less humped bill, squeaky rather than whistling calls, and the ranges of the two species overlap only slightly. Anis are very social, feeding and roosting in loose groups and even nesting co-operatively with several females laying eggs in the same nest.

Where to see It can be seen throughout the country up to about 2,300m but is largely replaced by the Smooth-billed Ani in southern Pacific areas.

Lesser Ground Cuckoo *Morococcyx erythropygius* 25cm

This slender, long-legged and long-tailed bird's most striking feature is its facial pattern. There is yellow in front of and around the eye and bare blue skin behind it, which is outlined in black, with a narrow white stripe above. The bill is substantial and slightly downcurved. The plumage is rufous below and grey-brown with a glossy sheen above. It forages mostly on the ground in search of invertebrates and prefers fairly open habitats such as savannah and woodland borders.

Where to see This cuckoo is resident on the north Pacific slope south to the Río Grande de Tárcoles and up to altitudes of about 1,200m.

Squirrel Cuckoo *Piaya cayana* 46cm

An unmistakable species with rich rufous plumage and a long, graduated tail with conspicuous white tips. It mostly forages in the forest canopy, on the forest edge and in scattered trees in open country. It may be seen running squirrel-like along branches, searching for small lizards and invertebrate prey, including snails, spiders, katydids and even caterpillars with stinging hairs or spines.

Where to see It is a common resident more or less throughout the country, from sea level to well over 2,000m.

Band-tailed Pigeon *Patagioenas fasciata* 35cm

This is the largest of the pigeons found in Costa Rica. It is mostly purplish grey with a conspicuous white collar and an iridescent green patch on the back of its neck. Its long, rounded tail has a single broad dark band, and its bill and legs are yellow. Other large pigeons are smaller, ruddier and lack the white collar and the yellow bill and legs. The flight of this species is strong and direct, and it is often seen high over the forest canopy. It gathers in small flocks and feeds on the acorns, fruits and seeds that can be found in highland forests.

Where to see This species is found throughout Costa Rica's highlands and is common to abundant in partly forested areas above about 900m.

Pale-vented Pigeon *Patagioenas cayennensis* 30cm

This is a fairly large pigeon. As its name suggests, it has a whitish lower belly and undertail-coverts. It also has a distinctive slate-grey head which contrasts with its rufous-purple mantle. The female is similar to the male but duller. The Pale-vented Pigeon generally avoids forest and is found on the forest edge and in isolated trees in pastures, coconut palms, mangroves and river margins. It eats berries of many trees and shrubs, notably *Trema* and melastomes.

Where to see It is a common to abundant resident in the higher parts of Costa Rica's mountains, down as low as 900m locally, perhaps seasonally. It was formerly more common, but the large flocks of the past have been greatly reduced by shooting.

Red-billed Pigeon *Patagioenas flavirostris* 30cm

This is a large pigeon with a bluish grey body and wings, maroon neck and shoulders, and red bill with a yellow tip. It is usually seen in pairs or small flocks in clearings in patchy forest, farmland with scattered trees, and sometimes mangroves. Listen for low cooing song: *whoo, whoop-a-whoo*. It forages in trees for berries, acorns and buds, and sometimes on the ground where it is liable to become a pest when consuming sprouting corn or sorghum.

Where to see This pigeon is resident in Costa Rica and fairly common in the dry north-west lowlands and Valle Central. It is uncommon in the Caribbean lowlands south to Puerto Limón, and occasional in the south Pacific lowlands. It ranges up to altitudes of 2,100m.

Short-billed Pigeon *Patagioenas nigrirostris* 26.5cm

This is a rather dark, uniform pigeon that is found in lowland rainforest. It is unpatterned and mainly dark purplish brown. The female is similar but duller than the male. The male's often heard song is a far-carrying *cu-coo k'coo*, sometimes rendered as *who-cooks-for you*, accented on the second and fourth syllables. The Short-billed Pigeon forages in the treetops, or sometimes on the ground, for berries of many sorts, including mistletoes.

Where to see It is a common to abundant resident in lowland rainforest, up to 1,100m on the Caribbean slope and 1,450m on the south Pacific slope. It does not occur in the dry north-west Pacific area.

Inca Dove *Columbina inca* 20cm

The Inca Dove is rather like a ground-dove with a long tail. It is a pale, greyish-sandy colour with a bold, black scaly pattern covering most of its body. It has white outer tail feathers and its wings show a rufous patch in flight. The male's song is an often repeated *cowl-coo*, often paraphrased as *no-hope*. The Inca Dove is found in open country with a few trees, scrubby second growth, gardens and roadsides. It is mainly terrestrial, walking around eating small seeds and grit.

Where to see This is an abundant resident throughout the north-west lowlands, and occurs south in smaller numbers along the coast to at least Quepos. It is also common throughout the Valle Central as far as Ochomogo.

Ruddy Ground-dove *Columbina talpacoti* 16.5cm

The male Ruddy Ground-dove is distinctive and well-named. The female is duller and browner but not dissimilar. Both sexes show rufous wing patches in flight. The Ruddy Ground-dove is a species of agricultural fields and semi-open areas with sparse vegetation, and also forages in gardens in rural villages. It often forms small flocks that forage on the ground for seeds and fallen berries, and may be seen on roads, apparently consuming grit to aid digestion.

Where to see It is resident and very common in deforested lowlands on the Caribbean slope and south Pacific slope. In some areas it is found as high as 1,200 or 1,400m. It is rare in Guanacaste except on the Pacific slope of the Nicoya Peninsula.

Ruddy Quail-dove *Geotrygon montana* 23cm

With its bright, ruddy colouration, the male Ruddy Quail-dove is very distinctive. The female is much browner but has a trace of the male's facial pattern and is still easily recognisable. Quite a shy species, it is usually seen when it is flushed explosively from a forest trail. It feeds on fallen fruits and seeds, and is said to visit manakin leks and fruiting trees to search for seeds regurgitated by other birds.

Where to see This is a fairly common resident species in the lowlands and foothills of the Caribbean slope, up to 1,000m, and on the south Pacific slope, up to 1,200m.

Olive-backed Quail-dove *Leptotrygon veraguensis* 21.5cm

This species is quite small for a quail-dove and much darker than most of its relatives. It is basically dark greyish with a contrasting white face and cheeks. The sexes are alike. The Olive-backed Quail-dove is found in dense rainforest in the lowlands and foothills where it forages for seeds, small fruits and some insects and worms. It is often seen walking along forest trails.

Where to see This quail-dove is a fairly common resident in rainforest in the Caribbean lowlands, north at least as far as Tortuguero. Locally, it is found up to about 450m in the foothills.

White-tipped Dove *Leptotila verreauxi* 26cm

This is an often solitary dove that has a brown back and wings, and pinkish-grey underparts extending up to its head. It has a narrow blue ring around the eye, which can be seen at close quarters. Particularly conspicuous are the broad white tips to its outer tail feathers, which are easily seen in flight. It is more evenly coloured than the Grey-headed Dove *L. plumbeiceps* (not illustrated) and has more white on its tail. It is largely terrestrial and inhabits open areas in deciduous woodland, plantations and gardens, where it searches the ground for seeds, small insects and grit, flying to cover when disturbed.

Where to see It is found throughout the Pacific slope up to about 1,000m and may be extending its range.

Grey-chested Dove *Leptotila cassinii* 24cm

A relatively dull, dark *Leptotila* dove, it is most easily identified by its rufous crown and nape. It is mainly terrestrial and can be found in the understorey of forest, often walking on trails, but it is more often seen in old second growth, coffee and cacao plantations, and shady gardens. It pushes aside dead, fallen leaves in search of seeds, grit and the occasional insect.

Where to see This is a common resident of both the Caribbean and south Pacific slopes, from sea level to 750m on the Caribbean slope and sometimes higher on the southern Pacific slope.

Mourning Dove *Zenaida macroura* 30cm

This is a slim species with a long, white-tipped tail. The wing-coverts and scapulars are boldly spotted with black. The female is browner overall. The Mourning Dove is a bird of open country with scattered trees, including lightly wooded areas and agricultural fields with hedges or sparse trees. It is also common in suburban areas. It forages singly or in flocks, gathering fallen seeds and occasional insects in stubble fields.

Where to see This dove is resident in the Valle Central and recorded occasionally in the Valle de General. It is also a common migrant that winters in Guanacaste, from about October to March. Occasionally, it turns up elsewhere.

Sungrebe *Heliornis fulica* 28cm

This attractive bird is ostensibly slightly smaller than the Pied-billed Grebe, though its longer wings and tail make it appear larger. Its boldly patterned black-and-white head and neck are distinctive. It lives on lakes, rivers and streams, usually in forested country, and feeds on insects, spiders, frogs and lizards. It rarely dives for food. When alarmed, it usually swims to cover but may take flight with shallow wingbeats.

Where to see This is a locally common resident, more or less throughout the Caribbean lowlands. It is not found in hilly terrain because it avoids fast-flowing water and rapids.

Grey-cowled Wood Rail *Aramides cajaneus* 38cm

This is a large rail that inhabits rainforest, wooded swamps and mangroves. Its grey, brown and chestnut plumage is made more distinctive by its coral-red legs, greenish-yellow bill and red eyes. Juveniles are duller. The loud, cackling call is a duet typically heard at dusk or at night. The Grey-cowled Wood Rail sometimes forages at night as well as by day, taking frogs, the occasional small snake, crabs and snails along with seeds and fallen fruits. It favours remnant forest patches and streamside woodland.

Where to see This rail is resident and locally common in suitable habitat from the lowlands to at least 1,500m.

Purple Gallinule *Porphyrio martinica* 33cm

With its violet-blue plumage and colourful red bill and yellow legs, this is an unmistakable species. It frequents ponds and marshes that have floating, emergent plants, enabling it to walk on the vegetation and swim only infrequently. Purple Gallinules eat the fruits of water lilies and other plants but also catch small fish, frogs and a variety of aquatic insects and other invertebrates. They have also been known to take the eggs and chicks of jacanas.

Where to see The Purple Gallinule is locally common countrywide, from sea level up to 1,500m, wherever there is suitable habitat.

Limpkin *Aramus guarauna* 66cm

This is a large wading bird with long neck and legs, and largely brown plumage. It has a long, laterally flattened bill and a characteristic halting walk, which suggests limping. Its favourite food is snails but it also takes other invertebrates as well as small reptiles and amphibians. It has a loud wailing call, most often heard in the evening and at night when it roosts in trees. It frequents river margins, wooded swamps and marshes.

Where to see Found chiefly in the Tempisque Basin and Caño Negro region. Now rare in other lowland areas on both the Pacific and Caribbean slopes.

Least Grebe *Tachybaptus dominicus* 24cm

This grebe is a very small, dark waterbird with bright yellow eyes and a slim, sharply pointed bill. In flight it shows a white patch on the secondaries. It is found on small ponds, lakes or even slow-flowing rivers. It dives for food, which includes small fish, frogs, crustaceans and aquatic insects.

Where to see It is resident on both slopes from the lowlands up to about 1,500m.

Pied-billed Grebe *Podilymbus podiceps* 33cm

This species is slightly larger than the Least Grebe and mainly dark brown. During the breeding season it has a black throat, a white eye-ring and a white bill with a black ring around it. It is found on freshwater ponds, lakes and slow-flowing rivers. Like other grebes, it dives for food but takes more fish than the Least Grebe. It has a loud, whinnying call.

Where to see This is a locally common resident in the lowlands and middle elevations on both slopes, though it is seen most frequently in Guanacaste. It moves frequently in response to changing water levels.

Double-striped Thick-knee *Burhinus bistriatus* 50cm

This is a large bird with long legs and an upright stance, a white supercilium bordered above by a black stripe, and big yellow eyes. Otherwise, it is a dull, grey-brown bird with streaks and bars that make it inconspicuous in the scrubby grassland and open woodland in which it occurs. It is most active after dusk, particularly on moonlit nights. Its loud calling at night is a characteristic sound of the Guanacaste dry forest.

Where to see Though it resembles a shorebird, it is found in savannah and scrubby woodland and is a species that has benefitted from forest clearance. It is a fairly common resident more or less throughout the lowlands of Guanacaste.

Northern Jacana *Jacana spinosa* 23–25.5cm

Jacanas are rail-like water birds but have long legs and very long toes that provide support, enabling them to walk on aquatic vegetation. At rest, adult Northern Jacanas appears mostly chestnut with a black head and breast. After alighting from flight, the wings are extended briefly, displaying the yellow flight feathers. Note the yellow spur on the wings. Immatures have a black crown, eye-stripe and back of neck, and the underparts are largely white.

Where to see This is an abundant resident on freshwater ponds and marshes with floating vegetation in Guanacaste. It also occurs locally in small numbers on both slopes up to 1,200m or more.

Juv

Black-necked Stilt *Himantopus mexicanus* 38cm

This is an elegant bird with a long, slender neck, long pink legs and a long bill. Adults are mainly black above and white below with a white central back, rump and tail that is conspicuous in flight. Their very long legs allow stilts to feed in deeper water than most other waders. Stilts are quite gregarious, sometimes gathering in loose groups of 50 or more. They forage in both saltwater and freshwater, searching for tiny aquatic invertebrates, including insects, crustaceans and molluscs.

Where to see This is a locally common resident, most likely to be seen around the Gulf of Nicoya and suitable habitat in Guanacaste, particularly in the Tempisque Basin. Sometimes seen in small numbers elsewhere in the lowlands.

Hudsonian Whimbrel *Numenius hudsonicus* 43cm

The Hudsonian Whimbrel is a mottled brown wader with a long, downcurved bill and conspicuous dark stripes on the crown and through the eye. It forages on mudflats, the edges of tidal creeks and occasionally on wet pastures. It feeds mainly on marine invertebrates, including crabs, molluscs and marine worms. It often associates with other waders when roosting. Its calls include a loud, melodious trilling.

Where to see This is a migrant and winter visitor that is most easily seen on the Pacific coast, especially around the Gulf of Nicoya. A few migrants also occur on the Caribbean coast. During the summer, a few non-breeding birds remain on both coasts.

Ruddy Turnstone *Arenaria interpres 22cm*

The breeding plumage of this wader (seen on the right of the photo) is rufous and black on the upperparts with a black chest and otherwise white underparts. The head has bold black markings and, in flight, a distinctive pied pattern on the wings. The bill appears slightly upturned at the tip. The bird on the left is still in winter plumage. Ruddy Turnstones forage mainly in intertidal zones, flicking aside pebbles and seaweed in search of molluscs and other invertebrates.

Where to see This species is a migrant and winter visitor to both coasts. A few remain as non-breeding summer visitors.

Spotted Sandpiper *Actitis macularius* 19cm

This is a rather nondescript wader in non-breeding plumage when it has brown upperparts and white underparts, yellow legs and a white supercilium with a dark eye-stripe. In summer breeding plumage the eye markings are more defined, there is dark barring on the back and bold spotting on the underparts. Its most characteristic feature is its forward-tilted gait and continuous bobbing of the rump and tail. It also has a distinctive jerky flight and a frequently used *peet-weet* call.

Where to see This is a widespread winter visitor that occurs in all waterside habitats up to 1,800m. It moults into its spotted summer plumage before migrating north in April/May.

Sunbittern *Eurypyga helias* 48cm

This beautiful bird is most easily seen in forested areas along fast-flowing, rocky rivers or small forest streams, or swampy pools. The black, red and yellow 'sunburst' pattern on its wings, used as a defensive display, is its most striking feature. Wading in shallow water, it forages for a great variety of streamside or aquatic prey, including lizards, frogs, crabs and other invertebrates.

Where to see Resident on the Caribbean and south Pacific slopes up to about 1,500m. Most readily found in forested foothills.

Laughing Gull *Leucophaeus atricilla* 40cm

In summer plumage, the Laughing Gull has a black head and bill, white eyelids, and a white neck and underparts. The immature lacks the well-defined black head but has some dark mottling. It is similar to the adult in non-breeding plumage, where the black hood is replaced with a black eye-mask. Laughing Gulls are named for their cackling call. They feed mostly in coastal areas, scavenging scraps from fishing boats, but will also catch small fish in shallow water.

Where to see This species is a migrant and winter resident. In summer, there are also a few non-breeding residents, mostly on the Pacific Coast in the Gulf of Nicoya.

Wood Stork *Mycteria americana* 102cm

This is a mainly white, long-legged wading bird with a black to dark grey, featherless head and neck, and a heavy bill that curves down slightly at the tip. Only when flying do the black flight feathers become fully visible. Like other storks, Wood Storks fly with their neck extended and legs trailing. This is a gregarious species that often hunts co-operatively, driving and concentrating fish to make them more easily caught. Wood Storks also roost communally and nest in colonies.

Where to see This stork is common to abundant in Guanacaste and around the Gulf of Nicoya. Elsewhere it is much less common.

Jabiru *Jabiru mycteria* 135cm

This is a huge stork with an all-white plumage. The head and neck are bare, the skin black except for a red collar at the base of the neck. The heavy bill is black and slightly upturned. Like other storks, it often soars. It forages in shallow freshwater marshes and lakes, preying on large fish, eels, frogs and the occasional snake, young bird or mammal.

Where to see This impressive stork is becoming less common in its only Costa Rican breeding area, the Tempisque Basin. It is seen only occasionally elsewhere in Costa Rica.

Magnificent Frigatebird *Fregata magnificens* 91–111cm

This is a large, aerial seabird, mostly blackish with a deeply forked tail, long hooked bill and a 2m wingspan. Females have white on the breast and neck; immatures are similar but also have white on their head. Males have pink bare throat patches, which are bright red during the breeding season and inflated during display. Although they are seabirds, they seldom enter the water, mostly snatching their prey from the surface and also harrying other seabirds.

Where to see This frigatebird occurs on both coasts but is commoner on the Pacific side, particularly in the Gulf of Nicoya. It breeds on offshore islands.

Anhinga *Anhinga anhinga* 86cm

A relative of cormorants, the Anhinga is easily identified by its elongated appearance with a small head and sharply pointed bill, long thin neck and long tail. When fishing, it often swims with only its head and neck showing and, underwater, it spears fish with a rapid thrust of its bill. It also eats small snakes, baby caimans and aquatic insects.

Where to see A widespread but mainly uncommon resident found throughout the lowlands of both slopes.

American White Ibis *Eudocimus albus* 63cm

The adult American White Ibis is white except for its black wing tips which are conspicuous in flight. Its long thin, downcurved bill is red (with a blackish tip during the breeding season). The bare facial skin and the legs are also red. Immatures are mostly brown with a white rump and belly; they lack the bright red face, though their legs are pinkish. The neck is long and extended in flight. This species is gregarious, and flocks forage in both fresh and salt water, probing mud for food.

Where to see This ibis occurs in the lowlands of both slopes but is most easily seen around the Gulf of Nicoya and in the Tempisque Basin.

Roseate Spoonbill *Platalea ajaja* 81cm

The Roseate Spoonbill's bright pink plumage and legs, and its long, spatulate bill, make it unmistakable. The head of the adult is bare and pale green while that of the immature is feathered and white. Roseate Spoonbills are gregarious and feed in the shallows of both fresh and salt water by swinging their bill from side to side, stirring up sand or mud, enabling them to detect their aquatic prey.

Where to see It can be found chiefly around the Gulf of Nicoya and less commonly elsewhere along the Pacific coast. It is rare on the Caribbean coast.

Rufescent Tiger Heron *Tigrisoma lineatum* 66cm

This is a handsome heron with a chestnut head and neck barred with black. It frequents streams, ponds and swamps inside forest, and is rarely seen in open habitats, at least in the daytime. It is usually solitary, standing motionless at the water's edge, waiting to ambush passing prey, including fish, frogs, large insects and even the occasional rodent or snake.

Where to see It is an uncommon to rare, localised resident in forest throughout the Caribbean slope, from sea level up to about 500m.

Bare-throated Tiger Heron *Tigrisoma mexicanum* 80cm

This is the largest of the three Costa Rican tiger herons and its bare yellow throat distinguishes it from the other two species. The Fasciated Tiger Heron *T. fasciatum* (not illustrated) is smaller and has a white throat. Unlike the Rufescent Tiger Heron, the Bare-throated Tiger Heron prefers larger, more open bodies of water, including lakesides, large ponds and marshes. It hunts at the water's edge, hunting fish, frogs, crabs and any other small prey within reach. Males have a loud booming call in the breeding season, mostly uttered at dusk or at night.

Where to see Frequents marshes and the banks of streams in the lowlands of both slopes but is most common and easy to see in the Tempisque Basin and elsewhere in Guanacaste.

Boat-billed Heron *Cochlearius cochlearius* 51cm

The large eyes and huge, shoe-like bill of this species are distinctive among herons. It is most easily seen along wooded riverbanks, the margins of lakes or in mangrove swamps. Boat-billed Herons forage mainly at night, a habit supported by their huge eyes. They are easily seen during the day at their communal roosting sites and at dusk when they disperse to begin foraging.

Where to see This species is most easily seen in the lowlands of the Caribbean and Pacific slopes, especially in Tortuguero National Park and the Tempisque Basin in Guanacaste.

Green Heron *Butorides virescens* 43cm

This is a small heron with yellow legs, a dull, velvet-green back, rich chestnut body and a dark cap that is often raised to form a short, untidy crest. The closely related Striated Heron *B. striata* (not illustrated) is sometimes considered to be the same species, but is rare in Costa Rica and can be differentiated by a grey or buff neck compared to the Green Heron's maroon. The Green Heron forages in shallow fresh water, usually close to cover. It feeds on small fish, frogs and aquatic insects.

Where to see Common and widespread throughout the country up to about 1,800m. The resident population is augmented by the arrival of winter visitors.

Western Cattle Egret *Bubulcus ibis* 51cm

This is a small egret, mainly white but tinged with buff on the head, breast and back and with a shortish, dagger-shaped, yellow bill. During the breeding season it has buffy plumes on its crown, back and chest and a hint of red on its facial skin and legs. It is usually seen in pastures and savannahs where it forages alongside grazing cattle and other livestock, feeding on grasshoppers and other insects. Western Cattle Egrets roost communally, often in huge numbers.

Where to see The Western Cattle Egret was first recorded in Costa Rica in 1954 but is now a common resident in deforested areas up to about 2,000m.

Br

Nbr

Great Egret *Ardea alba* 101cm

The Great Egret is the biggest, tallest and longest-necked of the white herons, often showing a kink in the neck. It has a yellow bill and black legs in all plumages. Great Egrets are often solitary and forage for fish, frogs or other prey in a wide range of both freshwater and saltwater habitats, including tidal flats, estuaries, marshes and river margins.

Where to see It is common and widespread in the lowlands of both the Pacific and Caribbean slopes. Most birds are winter visitors or migrants, though a few may be resident.

Tricolored Heron *Egretta tricolor* 66cm

This species is a slender, medium-sized heron, slaty-blue overall with a white belly and white underwings. It has a long neck and an extremely long bill. In breeding plumage, look for its yellow-orange facial skin and buffy plumes on the back. The Tricolored Heron frequents both freshwater and saltwater habitats, including estuaries, mangroves, mudflats, shallow wetlands and marshes. It is an active forager, darting around erratically as it chases after fish.

Where to see This attractive heron is a widespread but uncommon winter resident in the lowlands of both slopes, though very occasionally it occurs up to 1,500m. A few also occur as non-breeding summer residents.

Little Blue Heron *Egretta caerulea* 61 cm

This is a typical slim, long-necked heron with blue-grey plumage that darkens to purple on the neck and head. Juveniles are white except for dark wing tips which, along with yellow legs and stockier build, differentiate from Snowy Egret. They acquire adult plumage gradually so can appear patchy white and blue-grey at intermediate stages.

Where to see This heron feeds in both freshwater and saltwater habitats in the lowlands of both slopes. Most individuals are migrants or winter residents but some non-breeding birds remain for the summer and small numbers breed intermittently.

Snowy Egret *Egretta thula* 61cm

This white heron can be differentiated from the Great Egret by its smaller size, and yellow lores and feet. Unlike the juvenile Little Blue Heron, it has black legs and wholly white wings. In the breeding season it is adorned with lacy plumes on its head, breast and back. It can be seen in a variety of wetland habitats, especially river mouths, tidal mudflats, lagoons and shallow marshy pools. It usually forages fairly actively, often in groups, in shallow water, chasing after small fish or other small aquatic prey.

Where to see The Snowy Egret is a locally common winter resident in the lowlands of both slopes, sometimes up to 700m. It is also an uncommon to rare summer resident.

Brown Pelican *Pelecanus occidentalis* 109cm

This common pelican has an unmistakable long bill and unfeathered pouch. Its body plumage is dark brown with silvery frosting, and its head and neck are white, tinged yellow in the breeding season. It is often seen in small groups that fly in single file or V-formation, alternating flaps with long glides. It forages for fish by plunge-diving.

Where to see This is a common breeding resident on the Pacific coast and a widespread non-breeding visitor on the Caribbean coast.

King Vulture *Sarcoramphus papa* 81cm

This is a large vulture with striking black-and-white plumage and a bare, wattled head, brightly coloured orange, yellow and black. The all-dark juveniles are distinctly larger than Black and Turkey Vultures, and have a heavier bill and pale eye. In flight the juvenile lacks pale primaries and is uniformly dark. When soaring, both adults and juveniles hold their wings in a flat position. The King Vulture sometimes uses smell to find food under the canopy of rainforest but forages mainly in cattle country or other open areas.

Where to see Resident in small numbers throughout the country but the best areas to see King Vultures are probably Guanacaste and the Osa Peninsula.

Turkey Vulture *Cathartes aura* 76cm

The plumage of this vulture appears uniform black when at rest, though the head and cere are bare and red. In flight, the grey underside of the flight feathers is usually conspicuous. When gliding or soaring, the wings are held in a V-shape. It forages by flying fairly low and often homes in on gatherings of other vultures. It can locate food by smell so can even find small carcasses hidden below the forest canopy.

Where to see Common throughout the country, though rarely occurs above 2,000m. Usually in small numbers, except during migration, when large flocks move over the Caribbean lowlands. The resident population is augmented by migrants during the winter.

Black Vulture *Coragyps atratus* 64cm

This vulture is almost all black with an upright stance and a short tail. In flight, it shows broad wings with a pale patch on the underside of the primaries. It carries its wings horizontally when soaring and has characteristic quick wingbeats when flapping. Although not actively gregarious, several often occur together in thermals. It detects carcasses by sight, rather than by smell, and also homes in on groups of other feeding vultures.

Where to see It is a resident species though some of the population are thought to migrate. It occurs throughout Costa Rica, below about 2,000m.

Osprey *Pandion haliaetus* 58cm

This bird of prey has long, slightly pointed wings, which appear characteristically angled in flight and show dark patches at the bend on the underside. The upperparts are dark brown, contrasting with pale, almost white, underparts. A broad dark eyestripe shows clearly against the white of the head. The juvenile is very similar in appearance to the adult, but more mottled. Ospreys are most likely to be seen in association with large expanses of fresh or salt water, where they find their fish prey.

Where to see This is a migrant species, found in the lowlands and foothills of both coasts from September to April, though some birds remain all year round.

Hook-billed Kite *Chondrohierax uncinatus* 41cm

Its plumage is very variable, but this kite always has a longish tail and a heavy, deeply hooked bill. It has black, grey and rufous phases, though in all forms the tail is banded with pale grey and black. All phases also have a white eye, an orange spot on the lores and a greenish-yellow cere. The legs are yellow. This kite occurs in wooded swamps and secondary forest near water, where it hunts for land snails and apple snails (Ampullariidae).

Where to see This is a widespread but mostly uncommon species in the lowlands and foothills of both slopes.

Swallow-tailed Kite *Elanoides forficatus* 58cm

Its elegant shape and long, deeply forked tail make this an unmistakable raptor. It is usually seen in graceful flight or swooping swiftly to capture prey, including nestling birds, reptiles, frogs and large insects, which it sometimes eats in flight. It even takes fruit occasionally.

Where to see Mainly a breeding visitor from January or February until August or September, though a few are present year-round. It favours forested foothills and mountain slopes, mostly at 100–2,000m, occasionally higher. It avoids the dry north-west of Costa Rica.

Tiny Hawk *Microspizias superciliosus* 20cm

This is a small, thrush-sized hawk with orange eyes, and a yellow cere and legs. The male is dark grey above with a blackish crown and white below with fine dark barring. The female is similar but washed with brownish above and buff below. The Tiny Hawk occurs in humid forest and forest edge where it hunts for small birds, including hummingbirds which it ambushes at their leks.

Where to see It is an uncommon to rare resident in lowland and foothill forest (locally up to 1,200m) on the Caribbean slope and in the Golfo Dulce region on the southern Pacific slope.

Plumbeous Kite *Ictinia plumbea* 36cm

The Plumbeous Kite is a summer migrant to Costa Rica's tropical lowlands. It is a slender, long-winged raptor that is found in forest, including mangroves, and nearby semi-open areas with big trees. In flight, when hunting, the rufous patches on the underside of the primaries are very obvious. The Plumbeous Kite hunts by snatching small snakes, lizards and large insects from foliage. It also catches many insects in flight, especially bees and termites.

Where to see This kite is a breeding resident in the lowlands and foothills of both slopes. Costa Rican birds migrate south to South America in the non-breeding season.

Snail Kite *Rostrhamus sociabilis* 43cm

The Snail Kite is a unique snail-eating raptor that is widespread and common in Central and South America, including Costa Rica. It has a very sharply hooked bill, long broad wings, and short dark tail with a white base. The feet are orange. Females and immatures are streaky brown, males ashy grey. The Snail Kite feeds almost exclusively on snails found in freshwater marshes.

Where to see The Snail Kite is locally common in the Tempisque Basin and the Río Frío region. Otherwise, it is a rare and occasional visitor in the lowlands of both slopes.

Common Black Hawk *Buteogallus anthracinus* 56cm

This raptor is black except for the white band across its tail and a narrow white tail-tip. It also has a yellow cere and legs. In flight, the wings are very broad and rounded and the tail is short. This hawk tends to sit on low perches waiting to pounce on the crabs that are its staple prey. It also takes small reptiles, hatchling sea turtles and carrion. It is almost always associated with water, including mangroves, marshes, mudflats and rivers.

Where to see It is resident and locally common on both slopes up to 750m. Some authorities consider its population to comprise two species. Those in mangroves have shorter wings, and are sometimes split as the Mangrove Black Hawk *B. subtilis*.

Barred Hawk *Morphnarchus princeps* 61cm

This highland hawk has very broad wings and a rather short tail but its slaty-black head and throat, sharply contrasting with pale, barred underparts, are easy to see from a distance and probably its best identification features. The sexes are similar. The Barred Hawk mostly hunts from a perch inside forest or at the forest edge, from where it pounces on slow-moving prey, including snakes, frogs, crabs, large insects and occasional small mammals. Barred Hawks soar often, calling loudly.

Where to see This montane raptor is a fairly common resident in forested country on the Caribbean slope, mostly between 400m and 1,600m on the lower northern cordilleras and between 1,000m and 2,500m in the south.

Roadside Hawk *Rupornis magnirostris* 38cm

This is a small hawk with a grey head and chest, rufous barring on its belly and rufous primaries that are conspicuous in flight. It rarely soars, preferring to perch in open scrubby habitats and wait for prey (mainly small mammals, reptiles and large insects) to come within range. It is similar in general appearance to the Broad-winged Hawk *Buteo platypterus* (not illustrated) and Grey Hawk, but the Broad-winged Hawk is less grey and the Grey Hawk lacks rufous primaries.

Where to see This species is widespread but uncommon except in the lowlands of the north Pacific slope. On the Caribbean slope it is increasing and spreading into deforested areas.

White Hawk *Pseudastur albicollis* 56cm

This is an unmistakable, mainly white hawk with broad wings, black primaries and a short tail with a black subterminal band. It is often seen soaring above the forest canopy. Otherwise, it generally sits under the canopy waiting for prey to come within range. It pounces on small mammals, snakes, lizards, frogs and large invertebrates.

Where to see This hawk is a fairly common resident in hilly forested terrain from the lowlands up to about 1,500m on both slopes, though is absent from the dry Pacific north-west.

Semiplumbeous Hawk *Leucopternis semiplumbeus* 36cm

This hawk is dark slate-grey above and white below. Its black tail has a white subterminal band while its cere and legs are bright orange. It is a compact raptor with short, broad wings, well suited to its forest habitat where it is a 'perch-and-pounce' predator that hunts from perches in the sub-canopy. It feeds on small mammals, birds, lizards and snakes. The superficially similar Slaty-backed Forest Falcon *Micrastur mirandollei* (not illustrated) occurs in the same habitat but is larger with three white bands on its longer tail.

Where to see The Semiplumbeous Hawk inhabits humid forests, up to about 800m, in the lowlands and foothills of the Caribbean slope.

Grey Hawk *Buteo plagiatus* 41cm

This hawk is quite distinctive, being medium-sized and mostly pale grey with thin white bars on the underparts and a black tail with a clear white band. It often soars but usually not very high. It is found in partially forested areas or on the forest edge where it perches and waits for opportunities to ambush prey with sudden short flights, pursuing small mammals, occasional birds, small snakes, lizards and large insects.

Where to see It is a locally common resident in the north-west lowlands, south to the Río Grande de Tárcoles and locally up to 500m on adjacent slopes of hills. The Grey-lined Hawk *B. nitidus* (not illustrated) was split from the Grey Hawk and is very similar in appearance, but it inhabits south-western Costa Rica with little range overlap. It has a finely barred back and crown compared to the Grey Hawk's plain grey.

Short-tailed Hawk *Buteo brachyurus* 41cm

This is a rather small hawk that has two morphs. The pale morph (see photo) is brown, including its head and cheeks, but with white underparts. The dark morph is almost wholly dark brown, contrasting with the pale undersides to the outer flight feathers. It mostly occurs in forested areas. It is often seen soaring to great heights or swooping with great speed and agility over the forest canopy to snatch its prey, mostly small birds, snakes and lizards.

Where to see This hawk is an uncommon resident on the Caribbean slope of the Cordilleras de Tilarán and Central, and both slopes of the Cordillera de Guanacaste. Migrants from further north are recorded mainly in the Caribbean lowlands.

Zone-tailed Hawk *Buteo albonotatus* 53cm

The Zone-tailed Hawk is a dark, vulture-like raptor that flies with a pronounced dihedral to its wings. In fact, in flight it closely resembles a Turkey Vulture in its shape, colour and characteristic habit of rocking from side to side. The sexes are similar. It is found in open country with scattered trees, where its resemblance to a non-threatening Turkey Vulture enables it to get close to the small mammals, snakes and lizards upon which it preys.

Where to see The Zone-tailed Hawk is an uncommon resident in Guanacaste and on the lower slopes of the Cordilleras de Guanacaste and de Tilarán. Elsewhere, it is a migrant, sometimes seen in company with migrating Turkey Vultures or Swainson's Hawks *B. swainsoni* (not illustrated).

Ferruginous Pygmy Owl *Glaucidium brasilianum* 15cm

This is a diminutive owl with a relatively long tail and no ear tufts, which is often active during the day as well as night. It has two colour forms, one brown and the other more rufous, but both are white below with heavy streaking. It has two dark spots outlined in white on its nape that give the impression of false eyes. It hunts in wooded areas including plantations and suburbia. Its prey is mostly invertebrates but also some lizards. This owl is similar to the Central American Pygmy Owl *G. griseiceps* (not illustrated) of the Caribbean slope but their ranges do not overlap.

Where to see This species inhabits the lowlands of the north Pacific slope.

Tropical Screech Owl *Megascops choliba* 23cm

The Tropical Screech Owl has a facial disc that is more boldly outlined in black than other screech owls.. It also has fairly conspicuous ear-tufts. There is a rufous phase that is very rare. The sexes are alike. The male's song is a tremulous trill culminating in one or two loud hoots. The Tropical Screech Owl is found in open woodland, coffee plantations and even suburban areas with gardens and trees. It preys on large invertebrates, including moths taken at electric lights, spiders, scorpions and also sometimes small mammals, including bats.

Where to see This owl is resident in the highlands and foothills at altitudes of 400m to 1,500m, locally from the Cordillera de Tilarán south to Panama, mainly on the Pacific slope.

Pacific Screech Owl *Megascops cooperi* 23cm

This is a grey-brown owl with darker streaking. Its facial disc is outlined in brown, and it has yellow eyes and short ear-tufts. It is paler than the Tropical Screech Owl and its facial disc is less clearly defined. The Pacific Screech Owl lives in open woodland where it hunts at night, starting soon after dusk. It is usually detected by its gruff, bouncing-ball song. Its hunting strategy is to sit on low perches and pounce on large invertebrates.

Where to see This owl is a fairly common resident in the lowlands and foothills of the north Pacific slope up to 800m or more on the Cordillera de Guanacaste. It also occurs south of Carara, where it is found in mangroves.

Middle American Screech Owl *Megascops guatemalae* 20cm

This owl has softly dappled plumage that is less strongly marked with bars and streaks than that of other screech owls. It is smaller than its relatives and has a less well-marked facial disc. It has fairly inconspicuous ear-tufts. It is widespread (though not as common as the Tropical Screech Owl) in primary forest and old secondary growth. The song of the male is a soft, rather toad-like like trill. It mainly eats fairly large invertebrates, including katydids, beetles and spiders.

Where to see It is widespread but rare in some areas, locally common in others. It is resident in the wet lowland rainforest of the Caribbean slope and south Pacific slope. It ranges from sea level up to 1,000m.

Spectacled Owl *Pulsatrix perspicillata* 48cm

This is a large, distinctive owl. The spectacled appearance of adults makes them easily recognisable. Juveniles are equally distinctive, with a blackish facial disc and wholly white body. Adults have a bubbling call that has been likened to sheet metal being wobbled. Prey is very varied, including mammals, sometimes up to the size of skunks or agoutis, and roosting birds, including pigeons and motmots, as well as lizards, frogs and large insects. It is a bird of dense forest but often hunts in clearings.

Where to see The Spectacled Owl is widely distributed on both slopes up to about 1,500m.

Crested Owl *Lophostrix cristata* 40cm

This fairly large owl is easily recognised by its long white ear-tufts which form a striking white 'V' when the bird is alert. Its distinctive call is a gruff growl that is a characteristic noise in lowland rainforest at night. In spite of its size, this owl is said to prey mainly on large insects and sometimes a few small rodents and bats.

Where to see This is a widespread and fairly common resident on both slopes, from the lowlands up to about 1,500m.

Mottled Owl *Strix virgata* ♂28cm ♀35cm

This is a medium-sized owl without ear-tufts. Females are considerably larger than males. The facial disc is pale brown with whitish eyebrows, and the eyes are dark brown. Its calls include various hoots and a cat-like screech. It hunts in forest, semi-open areas and coffee plantations, taking rodents, small snakes and large invertebrates.

Where to see It is a fairly common resident in rainforest the length of both slopes from sea level to 1,500m. In the dry Guanacaste lowlands it is only found in evergreen forest bordering rivers.

Black-and-white Owl *Strix nigrolineata* 38cm

This owl has dark blackish upperparts, fine black-and-white barring on its underparts, and a yellow-orange bill and feet. It is strictly nocturnal and often attracts attention with its call – a gruff hoot, usually followed by two or three softer hoots. It often hunts at the forest edge, pouncing on rodents, large insects and even bats, the latter caught in flight.

Where to see An uncommon and local resident in the lowland rainforest and forested foothills of the Caribbean and north Pacific slopes, from sea level to 1,500m.

Resplendent Quetzal *Pharomachrus mocinno*

36cm (plus 64cm for the male's elongated tail coverts)

The male is the most ornate of the trogons, resplendent in its iridescent green and red plumage with elongated 'tail plumes'. The female is duller, with a shorter tail, though larger than other female trogons. The male's song consists of melodious, slurred notes. Also heard is a cackling call, like a demented chicken, uttered during a display flight above the forest canopy. Quetzals sally to pluck food in flight – fruit, particularly wild avocados, as well as lizards, frogs, insects and snails.

Where to see This spectacular bird is a fairly common resident in humid montane forests between 1,200m and 1,500m. It is most easily seen in Monteverde and on the Cerro de la Muerte.

♂

♀

♂

Slaty-tailed Trogon *Trogon massena* 30cm

The male is a large, glossy green and red trogon, best distinguished by the uniform blackish-grey underside of its tail and its orange-red bill. Any red plumage on females is confined to the lower belly and undertail. This trogon favours the canopy and middle levels of rainforest, where it forages for numerous fruits, as well as sallying for small lizards. It also follows monkey troops, on the lookout for flushed prey.

Where to see It is a common resident in rainforest in the lowlands and foothills on the Caribbean and south Pacific slopes, locally up to 1,200m.

♀

♂

Black-headed Trogon *Trogon melanocephalus* 27cm

This is one of the yellow-bellied trogons. The male has an iridescent green-blue back, rump and upper tail, with a blackish head and breast (slate grey in the female), unlike the Black-throated Trogon, which is largely green (male) or rusty brown (female). It lacks the finely barred undertail of the smaller Gartered Trogon, and the male Black-headed has a blue rather than yellow eye-ring.

Where to see This species is found in deciduous forest and tall secondary forest throughout most of the lowlands and foothills of the north Pacific slope, as far south east as Orotina and Tárcoles. It also occurs south of Lago de Nicaragua as far as the Río Frío district.

Baird's Trogon *Trogon bairdii* 28cm

The blue eye-ring of this species is diagnostic among the red-bellied trogons, and the lower breast, belly and undertail-coverts are a distinctive rich red-orange. The female is similar to the female Slaty-tailed Trogon but is less red below and has a bill that is bluish grey like the eye-ring. This species inhabits the upper levels of the rainforest where it feeds on fruits, lizards, frogs and many insects, usually snatched from foliage in flight.

Where to see Baird's Trogon is found in the lowlands and foothills of the south Pacific slope, north to Carara National Park, and locally up to 1,200m. It is a regional endemic with a range extending from Costa Rica into south-west Panama.

Collared Trogon *Trogon collaris* 25cm

Collared Trogons in Costa Rica have sometimes been regarded as a separate species, Orange-bellied Trogon (*T. aurantiiventris*). The orange to orange-red belly, separated from the breast by a white band, is distinguishing. Collared Trogons elsewhere have redder bellies but they are very similar; even their calls are the same. This species feeds on many different fruits and also takes many invertebrates – crickets, moths, caterpillars, beetles and spiders.

Where to see This trogon is a fairly common resident of rainforest and cloud forest in the middle and upper levels of all Costa Rica's cordilleras, from north to south. It occurs as high as 1,850m on the higher southern cordilleras.

Gartered Trogon *Trogon caligatus* 23cm

This species is one of the smaller yellow-bellied trogons. The male has a blue head and breast, a yellow eye-ring and black and white barring on the undertail feathers. The female has a dark grey head and back and an uneven greyish-white eye-ring. Gartered Trogons have the typical trogon foraging habits of perching, sallying and plucking fruits and insects while hovering. They prefer relatively open habitats like forest edges.

Where to see This species is resident countrywide in the humid lowlands, up to 830m or more on the Caribbean slope and higher, up to 1,200m, on the Pacific slope. It is uncommon in the dry north-west.

♂

Black-throated Trogon *Trogon rufus* 23cm

♀

This is a yellow-bellied trogon. The male differs from Gartered and Black-headed Trogons in its isolated black face in a metallic-green crown, breast and back, in its yellower bill, and from Gartered in its blue eye-ring. Duller than the male, the female is the only Costa Rican trogon to combine yellow underparts with a rusty-brown breast, head and back. Black-throated Trogons have the typical trogon feeding habit of taking both fruits and insect prey in flight.

Where to see This trogon frequents the low to middle levels of forest in the humid lowlands on the Caribbean and south Pacific slopes, sometimes as high as 1,000m.

♂

Amazon Kingfisher *Chlorocerlyle amazona* 29cm

This kingfisher is dark metallic green above with a white collar and white spotting on its wings and tail. It has a long, heavy bill and a prominent crest. The male's underparts are white with a rufous breast-band, while the female has green patches on the sides of its breast, with green spotting in the centre forming a broken band. The Amazon Kingfisher favours wide expanses of water, including large rivers, estuaries and mangrove channels where it dives for fish from a perch or sometimes while hovering.

Where to see This species is a common resident in the lowlands of both the Caribbean and Pacific slopes, occasionally as high as 900m.

♀

American Pygmy Kingfisher *Chloroceryle aenea* 13cm

This species is Costa Rica's smallest kingfisher. Males and females are similar except that females have a dark green breast-band across their rufous underparts. The American Pygmy Kingfisher frequents forest streams, shady lakes and ponds as well as narrow channels in mangroves. It hunts for prey from low perches, plunging for small fish and aquatic insects, and sallying for damselflies and other flying insects.

♂

Where to see This kingfisher is uncommon but widely distributed on both the Caribbean and Pacific slopes, occasionally up to 600m.

Green Kingfisher *Chloroceryle americana* 18cm

One of the smaller kingfishers, this species is green above and white below with green spotting down its sides. The male has a rufous breast-band while the female has two narrow green bands on the underparts. Both have white spots on the wings and outer tail feathers, which are conspicuous in flight. The Amazon Kingfisher is similar in appearance but is much larger and lacks the conspicuous white spotting on the wings and tail.

Where to see The Green Kingfisher forages in the small rivers, forest streams, pools and swamps that occur in the lowlands and mid-elevations of both slopes, up to 1,000m.

♂

♀

Ringed Kingfisher *Megaceryle torquata* 41cm

The Ringed Kingfisher is Costa Rica's largest kingfisher. It is rufous below and slate grey above, with a heavy dagger-shaped bill. The female has slate grey on the breast separated from the rufous by a narrow white band. This grey-blue breast is absent in the male but the sexes are otherwise similar. They have a bushy crest that can be more apparent than in the photographs. The Ringed Kingfisher forages in both fresh water and salt water, frequenting rivers, lakes and estuaries, where it plunges from higher perches than other kingfishers to catch good-sized fish.

Where to see This kingfisher is a fairly common resident in the lowlands of both slopes, sometimes up to 900m in foothill regions.

♀

♂

Lesson's Motmot *Momotus lessonii* 39cm

This species has mainly green upperparts, green-rufous underparts, a black mask and blue cap with black on the crown. It is confiding, often found close to houses, and attracts attention with its call, a soft double hoot, *hoop-hoop*. As with all motmots it perches in an upright position, often swinging its long tail back and forth. The long central feathers and their racquet tips are conspicuous but sometimes missing. It is found in various habitats, including rainforest, shady second growth and gardens.

Where to see Lesson's Motmots are found in the lowlands and foothills of the Pacific slope, up to 1,500m in the north and above 2,000m in the south.

Rufous Motmot *Baryphthengus martii* 46cm

This is the largest of the Costa Rican motmots. Its head, neck and breast are rufous and its back green, with blue on the wings that is visible in flight. It has a striking black mask and a small black chest-spot, and perches calmly for long periods, swinging its tail like a pendulum. Its diet consists of fruit, lizards, frogs and large insects that it consumes on a perch. It has a loud bubbling, hooting call which is heard most often at dawn.

Where to see The Rufous Motmot is common in rainforest in the lowlands and foothills of the Caribbean slope, up to 900m in the north and 1,400m in the south near Panama.

Turquoise-browed Motmot *Eumomota superciliosa* 34cm

This is a colourful motmot with a conspicuous pale turquoise stripe above the eye and longer shafts to the tail racquets than other motmots. It has a green and rufous body with a black mask and throat and turquoise-green wings. As with other motmots, it sits for long periods with an upright stance, occasionally sallying for passing insects or small lizards. It has a hoarse croaking call.

Where to see This motmot is a common resident in deciduous woodland and open scrubby areas of the north Pacific slope up to 800m. It also occurs in the Valle Central but is rare.

Broad-billed Motmot *Electron platyrhynchum* 30cm

This motmot is found in rainforest and attracts attention with its loud, harsh croaking call. It has the racket-tipped tail typical of motmots but the rackets are sometimes worn or missing. It is similar to the Rufous Motmot, but much smaller, with a blue chin, more restricted orange-rufous on underparts and a narrower black stripe through the eye. The broad bill is obvious if seen from below. It sallies to snatch prey from vegetation, including small lizards, frogs and many insects.

Where to see It is a fairly common resident in rainforest and old secondary growth on the Caribbean slope, from the lowlands up to 1,000m or more.

Rufous-tailed Jacamar *Galbula ruficauda* 23cm

This is an elegant bird with a slender body, and long bill and tail. It has glittering metallic-green upperparts and upper breast, and a rufous belly and undertail. The sexes are similar, differing mainly in the colour on the chin: white in males, buff in females. The song is a high-pitched, ascending trill. This species is usually encountered on the forest edge or in clearings, where it catches prey in flight – dragonflies, butterflies (including morphos) and other insects.

Where to see This jacamar is a fairly common resident in the lowlands and foothills of the Caribbean and south Pacific slopes, sometimes as high as 1,200m.

White-whiskered Puffbird *Malacoptila panamensis* 18cm

This is a stocky bird that often has its plumage fluffed up, giving the impression of a bird with a big head and short neck. Though the sexes are similar in pattern, males are much more cinnamon in colour, females more contrasting and greyer. This puffbird frequents the lower and mid-levels of forest and shady semi-open areas, sallying to the ground or vegetation to catch lizards, small frogs, insects and spiders. It sometimes joins mixed flocks of birds or follows swarms of army ants.

Where to see It is a fairly common resident in the lowlands and lower foothills (up to about 1,000m) on the Caribbean and south Pacific slopes.

Red-headed Barbet *Eubucco bourcieri* 15cm

This is a stocky, big-headed bird with a conspicuous yellow bill. The male, with its red head, throat and chest, is unmistakable, and the female is also very distinctive. The Red-headed Barbet feeds on many different fruits but also forages acrobatically in vine tangles, probing dead, rolled leaves for concealed crickets, moths and spiders, and gleaning foliage for insects. It often joins mixed-species flocks in the non-breeding season.

Where to see This barbet is resident at middle elevations on both slopes in cloud forest, secondary growth and semi-open thickets, mostly between 400m and 1,800m. It descends to lower elevations during the non-breeding season between October and February.

Prong-billed Barbet *Semnornis frantzii* 17cm

This is a rather stocky bird with a thick, silvery bill, tipped with a small hook. The sexes are alike except that males have a tuft of glossy black feathers on the nape. It is a territorial species while breeding, but forms flocks of a dozen or more at other times of the year and roosts communally in tree holes. It feeds mainly on fruits, but also takes petals, nectar and some insects.

Where to see This barbet is a fairly common resident in cloud forest at middle elevations, mainly 1,500–2,450m on both slopes but sometimes lower. It is endemic to Costa Rica and Panama.

Emerald Toucanet *Aulacorhynchus prasinus* 29cm

This is a small toucan with a mostly bright green body, a blue throat and a large black and yellow bill. It is typically seen in small loose flocks that forage in the forest canopy, on the forest edge, and in semi-open areas and clearings. Its loud call sounds rather like a saw cutting wood. It eats a great variety of fruits but also opportunistically takes insects, lizards and nestling birds.

Where to see It is a common or abundant resident of forested and semi-cleared areas at mid-elevations, 800–2,000m or above, highest on the higher cordilleras.

Fiery-billed Aracari *Pteroglossus frantzii* 43cm

This species is similar to the related Collared Aracari, though its bill is more brightly coloured orange-red with smaller serrations. It also has a brighter, mainly red band across its belly, and a larger breast spot. It is seen in small flocks of usually less than 12 birds, which forage at fairly high levels in humid forest for fruits, birds' eggs, nestlings, lizards and insects.

Where to see This aracari inhabits the humid forests of the south Pacific slope up to altitudes of 1,500m. Its range does not overlap that of the Collared Aracari. It is endemic to Costa Rica and Panama.

Collared Aracari *Pteroglossus torquatus* 41cm

This aracari has one of the less brightly coloured bills of the toucan family, but it does have conspicuous serrations on the upper mandible. Its plumage is mainly glossy-black above with yellow underparts, suffused with red, with a dark band across the upper belly and a black spot on the breast. The Collared Aracari is usually seen in small, straggling flocks in forest or secondary growth, foraging for fruits, birds' eggs and nestlings, lizards and insects.

Where to see Common in the lowlands and foothills of the Caribbean slope, generally below 1,000m, but rarely encountered on the Pacific slope. It is replaced by the Fiery-billed Aracari on the south Pacific slope.

Yellow-throated Toucan *Ramphastos ambiguus* 52cm

This is the largest of the Costa Rican toucans, easily distinguished by its yellow and dull maroon bill. Its shrill yelping call is conspicuous and often heard, particularly in the evening. Small flocks forage together, eating many different fruits and opportunistically taking nestling birds, occasional snakes, lizards and large insects.

Where to see This toucan is a common resident in the forested lowlands and foothills of the Caribbean slope and southern Pacific slope north to Carara National Park. It occurs as high as 1,200m, higher on the Cordillera de Talamanca.

Keel-billed Toucan *Ramphastos sulphuratus* 47cm

This large toucan has a flamboyant multicoloured bill and a croaking, frog-like call that is a characteristic sound of lowland rainforest. Keel-billed Toucans roam around the forest in small flocks, foraging for diverse fruits, large and small, as well as large insects and occasional small lizards and frogs.

Where to see This species is a common resident of rainforest and semi-open areas in the Caribbean lowlands, up to about 1,200m. It also occurs in the northern half of the Pacific slope.

Acorn Woodpecker *Melanerpes formicivorus* 21cm

This woodpecker has a distinctive black, white and red head pattern. The sexes are similar except for the female having a black forehead. This is a social species, seen in groups of 3–6, and well-known for storing acorns in natural crevices in tree trunks or in specially drilled holes. It also catches insects in flight.

Where to see This is a highland species, resident in forest and clearings with scattered trees, from about 1,500m up to the timberline on the Central and Talamanca Cordilleras.

Black-cheeked Woodpecker *Melanerpes pucherani* 18.5cm

This woodpecker has a heavily barred black back and white rump. The male has a conspicuous red crown and nape while the female has a white forehead and a black centre to its crown. The Black-cheeked Woodpecker is a bird of humid forests, favouring the higher levels of trees while foraging for insects and small fruits. It is very similar to the Golden-naped Woodpecker *M. chrysauchen* (not illustrated), which replaces it on the south Pacific slope and has a distinctive yellow nape.

Where to see It occurs commonly in the lowlands on the Caribbean slope, but only rarely on the Pacific slope.

Hoffmann's Woodpecker *Melanerpes hoffmannii* 18cm

This species is heavily barred on the back and pale buffy-brown below with a yellow belly. The male has a red crown and yellow nape. The female is similar but lacks the red crown and has a smaller area of yellow on its nape. It inhabits less dense woodlands, including secondary forest, plantations and gardens. It supplements its insect diet with fruits and nectar. It is similar to the Red-crowned Woodpecker *M rubricapillus (not illustrated)*, which replaces it on the south Pacific slope.

Where to see This species is common as high as 2,150m on the north Pacific slope. As forests are cut, it is also expanding its range on the Caribbean slope.

Hairy Woodpecker *Leuconotopicus villosus* 17cm

This is one of the smaller woodpeckers. The sexes are similar except that the male has a red patch on its nape which the female lacks. Otherwise, both sexes have a broad conspicuous white stripe down the middle of their black back and brownish underparts. The head is mostly black with white stripes above and below the eyes. This woodpecker forages for crickets, beetles and spiders at mid-levels in humid highland forests, and sometimes also at lower levels and even on the ground.

Where to see The Hairy Woodpecker is a common resident of the forested highlands, particularly oak woodland, from 1,500m up to the tree line. It ranges from the Cordillera de Tilarán south to Panama.

Rufous-winged Woodpecker *Piculus simplex* 18cm

This woodpecker is easily distinguished from the Golden-olive Woodpecker by its head pattern (lacking pale cheeks), blue eyes and rufous primaries. The female has a red hind neck but lacks the red crown and moustachial stripe of the male. Found in the canopy and edges of rainforest, where it forages for beetles, ants and other insects. It sometimes occurs in mixed-species flocks.

Where to see This is a fairly common resident of the forested lowlands and foothills of the Caribbean slope up to 750m and on the south Pacific slope to 900m.

Golden-olive Woodpecker *Colaptes rubiginosus* 20cm

This woodpecker has largely olive-green upperparts and heavily barred underparts. The male is distinguished by its broad red lateral crown-stripes, which meet on its nape, and its red moustachial stripe. The female has only a red nape. The most common call is a loud rattling trill. It mostly forages high in trees, searching for beetles, their larvae, termites and ants.

Where to see This woodpecker is resident on the forest edge, in semi-open woodland and in shady gardens. It is found at middle elevations, about 750–2,100m, on both slopes.

Chestnut-coloured Woodpecker *Celeus castaneus* 23cm

With large crest and chestnut plumage, both sexes of this woodpecker are easily identified. The male has a red cheek-patch, missing in the female. It is easiest to see in the canopy and subcanopy of dense rainforest or on the forest edge. It eats mostly ants and termites, pecking into termite tunnels and *Cecropia* trees for ants. It also removes flaking bark, eating whatever invertebrates are uncovered.

Where to see This woodpecker is an uncommon resident more or less throughout the Caribbean lowlands and locally reaching as high as 750m in foothill forest.

Lineated Woodpecker *Dryocopus lineatus* 33cm

This is a slimmer bird than the Pale-billed Woodpecker, with a white stripe across its face. The female has a black forehead and moustachial stripe, both of which are red in males. This woodpecker is rarely found in closed rainforest, preferring semi-open habitats, including trees in pastures and gardens. It forages in rotten tree trunks and branches for beetle larvae, and often raids ants' nests.

Where to see This is a fairly common resident in the lowlands on both slopes, ascending in semi-cleared areas to about 1,100m.

Pale-billed Woodpecker *Campephilus guatemalensis* 37cm.

This is a large, robust woodpecker with a bushy red crest and ivory bill. The bill differs from all Costa Rican woodpeckers bar Lineated in the north-west of the country, but the Pale-billed lacks the Lineated's black and white on face. The female Pale-billed Woodpecker is similar to the male but has a black throat and front to its crest. The drumming of this woodpecker is diagnostic – two loud taps in quick succession. It forages mainly in rotting wood, digging and probing for wood-boring beetles and their larvae.

Where to see This woodpecker is common in forested or partially forested areas on both slopes up to about 1,000m, higher on the south Pacific slope. It has declined in many deforested areas.

Crested Caracara *Caracara plancus* 61cm

This is a handsome bird with a short black crest, red facial skin and yellow legs. In flight it shows a large white patch on its primaries. Immature birds are duller and browner. The Crested Caracara competes with vultures for carrion and road kills. It also hunts for live prey, including rodents, nestling birds and small snakes.

Where to see It is a common resident in the dry north-west of the country, up to about 1,200m, and is increasing further south along the Pacific slope.

Laughing Falcon *Herpetotheres cachinnans* 53cm

This falcon has a white body but black wings and a striking black mask that extends round to the back of its head. It has a long black tail with several whitish bands. It tends to sit and wait on a high, exposed perch, ready to drop onto prey. Its main food is snakes, which it usually kills with a bite behind the head before carrying them off to a perch to eat. It is named for its cackling call.

Where to see It inhabits areas of open savannah, secondary growth and forest edges on both lowland slopes.

Barred Forest Falcon *Micrastur ruficollis* ♂33cm ♀38cm

This is a small bird of prey with short wings. Males have dark grey upperparts and white underparts, finely barred blackish. Females are dark brown above, and both sexes have a longish black tail with three narrow white bars. It hunts in the understorey of humid forest, feeding on small vertebrates and large insects.

Where to see It is a fairly common resident in forested regions of the Caribbean slope and south Pacific slope.

♂

Orange-chinned Parakeet *Brotogeris jugularis* 18cm

This small parakeet is green overall with bronzy shoulders and a small orange chin patch that is very hard to see. It also has a patch of yellow on the underwing, which is visible in flight, and a fairly short, pointed tail. It favours forest patches and open country with hedges and tall trees, including towns and villages, and is usually seen in pairs or small noisy flocks, foraging at flowering and fruiting trees.

Where to see It is a common resident in the north Pacific lowlands, up to about 1,200m, but is expanding its range following deforestation so can now be seen in most parts of the country.

White-fronted Amazon *Amazona albifrons* 25cm

This is a rather small *Amazona* parrot, easily identified by its obvious white forehead, blue crown, red on its face (extending behind eye) and yellow bill. In flight, the male shows a red patch on the front edge of its wings, not on the rear like both sexes of Red-lored Amazon. It is found in the canopy of both deciduous and evergreen forest but also in semi-open areas with scattered trees and forest patches, in both drier and more humid areas, including savannah and agricultural areas. It eats figs, *Inga* seeds and a great variety of other fruits, seeds and crops.

Where to see This parrot is a common to abundant resident in the lowlands and foothills of the Pacific north-west, including the Nicoya Peninsula and on the Pacific slope of the Cordillera de Guanacaste.

Red-lored Amazon *Amazona autumnalis* 34cm

Like other species of *Amazona*, the Red-lored Amazon is mostly green, but is distinguished by having red lores, yellowish-green cheeks and a lavender tinge to the crown and nape. The red patch on the rear of the wings is visible in flight but appears as only a small bar when at rest. The red on the face is characteristic of this species and is less extensive than White-fronted Amazon, being restricted to the lores and forehead. It is a parrot of forest edge and semi-cleared areas. Like most *Amazona* species, the Red-lored Amazon is gregarious and noisy, with a rather metallic call.

Where to see The Red-lored Amazon is common to abundant in partly forested areas in the lowlands and foothills of the Caribbean slope up to about 800m; and also on the south Pacific slope up to about 1,000m.

Mealy Amazon *Amazona farinosa* 38cm

The Mealy Amazon is a large, green parrot that lacks the colourful head markings of most species of *Amazona*. Its best field mark is the pale eye-ring on its otherwise rather plain head. It does, however, have the red patch on the rear wing, like Red-lored Amazon. The Mealy Amazon is generally encountered in heavily forested areas, where it keeps to the canopy, but also descends to gaps and visits the forest edge to feed. Its diet includes the fruits, seeds and flowers of many forest trees.

Where to see The Mealy Amazon is a common to abundant resident of forested areas of the Caribbean and south Pacific lowlands, usually no higher than 500m. Numbers decline sharply in deforested areas.

Orange-fronted Parakeet *Eupsittula canicularis* 22.5cm

Parakeets in the genus *Eupsittula* are small with relatively long and somewhat pointed tails. The Orange-fronted Parakeet's head is distinctive, with a blue crown, orange forehead, bare yellow eye-ring, yellow iris and white bill. It is common in the forest canopy as well as semi-open areas with scattered trees, plantations, and locally in towns and suburban areas. It is often in flocks which sometimes number 30 or more outside the breeding season. It eats flowers (*Gliricidia*), fruits (*Ficus*, *Bursera*) and seeds (*Ceiba*, *Inga*).

Where to see The Orange-fronted Parakeet is a common resident of the north Pacific lowlands, up to 1,000m on the slopes of the Cordilleras de Guanacaste and de Tilarán, and on the hills of the Nicoya Peninsula. Populations of this species have decreased in recent years due to trapping for the pet trade.

Scarlet Macaw *Ara macao* 84cm

This is an unmistakable macaw with a large, powerful bill, spectacularly colourful plumage, a long tail and a raucous voice. It is usually seen in pairs, family groups or small flocks (sometimes up to 25 birds) flying over rainforest or foraging in the canopy for fruits and nuts. Up to 50 or more may gather in communal roosts at night.

Where to see This macaw is an inhabitant of lowland rainforest, found locally up to altitudes of 800m. It was once abundant throughout the lowlands but now good numbers are found only on the Pacific slope in Carara National Park and on the Osa Peninsula, particularly in Corcovado National Park.

Tawny-throated Leaftosser *Sclerurus mexicanus* 15cm

This leaftosser differs from the Grey-throated mainly in the colour of its throat – rufous rather than grey. Both are ground-dwelling birds that forage for cockroaches, beetles and spiders on forest trails and the forest floor, tossing aside leaf litter and probing damp soil and dead wood in their search for grubs, worms and other invertebrates.

Where to see This species is an uncommon resident of middle elevations on both slopes, from 700m up to 1,500m on the Caribbean slope, and from 1,000m up to 1,850m on the Pacific slope.

Grey-throated Leaftosser *Sclerurus albigularis* 17cm

This leaftosser has a dark brown back and wings, a black tail, a chestnut chest and rump and a grey throat. Its earthy colours make it rather inconspicuous, though it is not particularly shy. It is often seen on forest trails at dawn and dusk. The grey throat-patch distinguishes this species from the otherwise similar Tawny-throated Leaftosser.

Where to see Most likely to be found at mid-elevations on the Caribbean slope but its range spills over to the Pacific slope of the northern cordilleras, where it is often seen at Monteverde.

Olivaceous Woodcreeper *Sittasomus griseicapillus* 15cm

This is a small, unstreaked woodcreeper with a short, slender bill. Its head, neck and underparts are mainly greyish while its wings, rump and tail are chestnut-rufous. It is a species of rainforest and old second growth in humid areas but is also found in more open woodland in drier areas. Climbs actively on trunks and branches, often high up, searching for insects and spiders in crevices and under loose bark. It has a characteristic, rapid trill that often attracts attention.

Where to see This species is a common resident in the foothills and highlands, between 500m and 1,600m, throughout most of the country. It is an uncommon resident in the north-west lowlands.

Plain-brown Woodcreeper *Dendrocincla fuliginosa* 21cm

As its name suggests, this woodcreeper is relatively uniform in colour and lacks much in the way of streaks or spots. It does, however, have a fairly obvious dark malar stripe on its otherwise pale face. Like all woodcreepers, it also has stiff, spiny tail feathers to support it on tree trunks. The Plain-brown Woodcreeper is a regular follower of army ants, dropping down from perches above the swarm to seize fleeing insects and spiders. Its song is a shortish, indistinct rattle.

Where to see This woodcreeper is fairly common in the Caribbean lowlands and foothills up to about 750m. Through low passes, it also reaches the Pacific slope of the Cordillera de Guanacaste.

Wedge-billed Woodcreeper *Glyphorhynchus spirurus* 15cm

This is a small, slender woodcreeper that forages over trunks and branches in search of prey. It has a short, laterally flattened and slightly upturned bill to facilitate probing for food items in bark and epiphytes, mostly small insects, spiders and other invertebrates. Its brown mottled colouring makes it inconspicuous, particularly when it remains still. Its chisel-shaped bill is diagnostic. The song is a high trill that gets louder before stopping suddenly.

Where to see This species is common in the lowlands and foothills of the Caribbean slope and the southern Pacific slope north to about Carara, locally as high as 1,500m.

Strong-billed Woodcreeper *Xiphocolaptes promeropirhynchus* 30cm

This is a large woodcreeper with a hefty, slightly downcurved bill and heavily streaked head, with streaks extending to the breast and belly. It is found in dense rainforest, often in pairs or with a mixed-species flock, foraging for prey at all levels, from near the ground up to mid levels in the forest. It is also encountered at army-ant swarms. It calls often – a loud *cooo-weeew*.

Where to see This is a fairly rare resident woodcreeper, often inconspicuous, occurring in lower-middle elevations (500–1,700m) from the Caribbean slope of the Cordillera de Tilarán south to Panama.

Northern Barred Woodcreeper *Dendrocolaptes sanctithomae* 28cm

This is one of the larger woodcreepers and is uniformly barred over most of its body. It has a substantial dark bill, and forages at low and middle levels on tree trunks and branches in rainforest, prying off bark to expose prey. It often attends army-ant swarms, perching above the swarm and dropping down to catch insects, spiders and occasional lizards and frogs. Its song is a series of bisyllabic whistles.

Where to see This species is a common resident in rainforest from the lowlands up to mid-elevations, locally up to 1,300m on both slopes. It is rarer in the dry areas of the north-west.

Cocoa Woodcreeper *Xiphorhynchus susurrans* 21.5cm

This is a typical woodcreeper with rufous wings and tail, a buff body with darker streaks, and a pale stripe behind the eye. Its unmarked buff throat and dark upper mandible on strong, slightly downcurved bill differentiate it from other woodcreepers. It forages for small invertebrates by probing bark and epiphytes as it ascends trunks and searches branches, starting from a low level. Generally seen alone but groups may attend army-ant swarms. Its song is a long series of upwards-inflected whistles.

Where to see This woodcreeper is widespread, occurring in humid forest edges and clearings in the lowlands and foothills of the Caribbean slope up to around 650m. It occurs slightly higher on the south Pacific slope, up to 900m.

Black-striped Woodcreeper *Xiphorhynchus lachrymosus* 24cm

This distinctive woodcreeper is more boldly patterned than most. It has plain rufous flight feathers and tail but the rest of its plumage has bold black edging and white streaking, giving a spotted and streaked effect. Its bill is straight, laterally compressed and substantial. As is characteristic of woodcreepers, it climbs up tree trunks and large branches searching for small lizards and diverse invertebrates. It often joins mixed flocks when foraging. Listen for its song – a clear descending series of notes, often stuttered at the beginning.

Where to see It is a common resident of rainforest in the lowlands of the Caribbean slope and south Pacific slope as far north as Carara. Locally, it occurs up to about 1,000m, a little higher on the Pacific slope.

Spotted Woodcreeper *Xiphorhynchus erythropygius* 23cm

This species has the rufous wings and tail common to most woodcreepers but the rest of its plumage is olive-brown with a distinctive spotted pattern, particularly on the underparts. The broad pale 'spectacles' are distinctive. The bill is straight, long and darker on the upper mandible than the lower. It forages on tree trunks and branches, where it can be acrobatic, often clinging upside-down, in its search for invertebrates and other small prey, sometimes including frogs and salamanders. It has a melancholy-sounding song of two or three clear whistles.

Where to see It inhabits humid rainforest at mid-elevations, from 700m up to 1,450m on both slopes and sometimes higher.

Brown-billed Scythebill *Campylorhamphus pusillus* 23cm

This is a strongly streaked woodcreeper, easily distinguished from others by its very long, slender, downcurved bill. It has a long complex song of clear trills and whistles that often sounds as if two songs were being uttered at the same time. It lives in dense wet forest, mostly in hilly terrain, and probes tree crevices for beetles, insect larvae, spiders and other invertebrates. Often accompanies mixed-species flocks.

Where to see It is an uncommon resident of the middle elevations of rainforest, at 300–1,500m, on the Caribbean slope.

Streak-headed Woodcreeper *Lepidocolaptes souleyetii* 19cm

This is one of the medium-sized woodcreepers and it has a slightly downcurved, usually pinkish bill. It has the usual rufous wings and tail, a buff throat, and buff streaking edged with black on the underparts and upper back. The top of its head has fine buff streaks. It clings to bark in the usual woodcreeper fashion and hitches up tree trunks and along branches, prying off bark to expose invertebrate prey, including moths, beetles and spiders. Its song is a rapid descending rattle.

Where to see It is found in woodland, plantations and areas with scattered trees in the lowlands of both slopes, regularly up to 1,500m, but it is rarer in the dry north-west.

Plain Xenops *Xenops minutus* 12cm

This acrobatic forager is small and brown but with a conspicuous silvery-white cheek-stripe and a less distinct pale supercilium. The lower mandible of its bill is curved upwards, giving the whole bill the appearance of being tilted upwards. It is more uniform than the Streaked Xenops *X. rutilans* (not illustrated) but their ranges rarely overlap. The Plain Xenops forages in the lower and mid-levels of trees and hangs tit-like on decaying twigs, probing dead leaves in search of its invertebrate prey. It often joins mixed-species flocks.

Where to see It inhabits the lowlands and foothills on both slopes, from sea level to 1,100m or more, but is rare in the drier regions of the north-west.

Scaly-throated Foliage-gleaner *Anabacerthia variegaticeps* 16cm

With its conspicuous orange 'spectacles', this species is one of the easier foliage-gleaners to identify. It is an active bird that forages at all levels, sometimes hanging upside-down to search through mossy tangles and epiphytes. It is often seen in pairs or family groups and also joins mixed-species feeding flocks. It takes many different prey items, including katydids, moths, beetles and spiders. It also flutters after escaping prey, catching it in flight.

Where to see This species is a resident of humid rainforest at middle elevations, from 800m to 1,850m, the length of the Caribbean slope and on the Pacific slope of the Cordillera de Talamanca.

Lineated Foliage-gleaner *Syndactyla subalaris* 19cm

This foliage-gleaner is basically rufous-brown but has more obvious buff streaking all over its head, back and underparts than most other relatives. It also has a bright rufous tail. The Lineated Foliage-gleaner forages in the forest understorey, usually in a mixed-species flock, searching vine tangles and clusters of dead leaves for small prey, including crickets, cockroaches, beetles, moths and spiders. It also takes occasional small lizards and frogs.

Where to see This bird is a common resident in rainforest on both slopes, mainly between 600m and 2,300m on the Caribbean slope, and 1,000m and 2,300m on the Pacific slope.

Streak-breasted Treehunter *Thripadectes rufobrunneus* 21.5cm

This species is largely rufous-brown but lacks streaking on its back. It has a distinctive buff throat and buff streaking on its breast. It also has a relatively heavy, dark bill. It forages actively, poking around in epiphytes and tangled vegetation in the heavy undergrowth of highland forests and feeds on invertebrates as well as occasionally taking small reptiles and amphibians.

Where to see It is found in cool, wet cloud forest and occurs on both slopes of most cordilleras throughout Costa Rica, mostly between 800m and 2,500m, sometimes higher.

Fawn-throated Foliage-gleaner *Automolus cervinigularis* 18.5cm

This species has olive-brown upperparts, plain yellow-brown underparts and a rather indistinct buffy forehead, extending back to a stripe over and around the eye. It has fairly uniform colouring though its throat is distinctly paler, sometimes almost white. It forages in the lower levels of humid rainforest, including old secondary growth and plantations, occasionally descending to the ground. It feeds mainly on invertebrates but also takes a few small frogs and reptiles.

Where to see Occurs in the lowlands and foothills, up to 1,200m, on the Caribbean and south Pacific slopes.

Spotted Barbtail *Premnoplex brunnescens* 14cm

This small ovenbird is dark brown above with numerous buff, tear-shaped spots on its underparts. It inhabits wet rainforest and cloud forest, mostly in the understorey and often in mixed-species flocks with antwrens and warblers. It forages for small insects and spiders, searching through moss and epiphytes or probing crevices in and under bark.

Where to see This is a common resident in the middle elevations of both slopes, from 600m on the Caribbean slope, and 1,000m on the Pacific slope, up to about 2,450m.

Ruddy Treerunner *Margarornis rubiginosus* 16cm

This species is reddish brown with a conspicuous white throat and supercilium and small buffy-white spots on its breast. Sometimes it is seen alone or in pairs but more often it joins mixed-species flocks. When foraging, it searches mossy branches, twigs, dead leaves and epiphytes, probing for insects, spiders and other small prey items.

Where to see It is a common resident up to the timberline in montane forest the length of the country, Found in forest and forest edge upwards from 1,200–1,500m on the lower cordilleras and from 1,850m upwards on the Cordilleras Central and de Talamanca.

Slaty Spinetail *Synallaxis brachyura* 15cm

The Slaty Spinetail is a slender bird with a long, pointed, wispy tail. Its plumage is mainly dark grey-brown, becoming dark grey on the head. Its crown and shoulder patches are rich rufous. The Slaty Spinetail can be difficult to see while foraging in tangled thickets for caterpillars, beetles and other invertebrate prey, but its hard, low-pitched, grating call often attracts attention.

Where to see This species is a widespread and common resident in scrubby habitats and overgrown pasture on both slopes, up to about 1,500m on the Caribbean slope and up to 1,250m on the Pacific slope south of Carara.

Dot-winged Antwren *Microrhopias quixensis* 11cm

Both sexes of this small, warbler-like antwren are distinctive. The male is velvety black overall while the female is slaty grey above and rich rusty below. Both sexes have a white wing-bar and spots on their wings, plus a white-tipped tail. The Dot-winged Antwren is often seen in pairs and sometimes in mixed-species flocks. It feeds on small insects and other invertebrates, busily searching through foliage, leafy thickets and vine tangles for its prey.

Where to see It is a common resident in humid forest and forest edge throughout the lowlands of the Caribbean slope and Pacific slope south of Carara, locally as high as 1,000m.

♀

♂

Slaty Antwren *Myrmotherula schisticolor* 10cm

The male is dark slate-grey with a black throat and breast, plus narrow white wing-bars. The female lacks the wing-bars, and is olive-brown above and a paler and brighter brown below. Males and females often forage together in the understorey of rainforest and will join mixed-species flocks. They forage busily, gleaning small insects and spiders from foliage and rolled leaves.

Where to see It is resident in rainforest at mid-elevations on both slopes, upwards from 700m in the north and 1,000m in the south-east, to as high as 1,700m on the Caribbean slope and 1,500m in the Pacific north-west. It also occurs at sea level in the Golfo Dulce region.

Plain Antvireo *Dysithamnus mentalis* 11.5cm

The male is largely grey with a darker head, narrow white wing-bars and paler underparts, becoming pale yellow on the belly and undertail. The female is olive brown above, with buffy-yellow underparts (palest at the throat), a rufous crown and pale eye-ring, but lacks white wing-bars. This is a busy bird, usually seen in pairs or family groups, which forages in the understorey of humid forests, gleaning invertebrates.

Where to see It inhabits the foothills of the Caribbean slope and the south Pacific slope.

Barred Antshrike *Thamnophilus doliatus* 16cm

This species is smaller than the Fasciated Antshrike (whose male is similarly barred), with a bushier crest and pale yellow eyes. The female also differs in having rufous upperparts and buff underparts. The Barred Antshrike is typically seen in low scrubby thickets, on the forest edge and in low tangled vegetation, often attracting attention with its distinctive song: an accelerating series of chuckling notes. It forages for the usual antshrike fare of diverse invertebrates, including grasshoppers, caterpillars, beetles and spiders.

Where to see This species is resident in the lowlands and foothills of both slopes, up to about 1,400m. It is most common in Guanacaste in the dry north-west of Costa Rica.

Black-hooded Antshrike *Thamnophilus bridgesi* 16.5cm

The male is almost entirely black, though it fades to slate-grey on the lower underparts and there are small white spots on its wings and the sides of the back. The female is more olive-grey and streaked on the head, throat and breast. Males and females often forage together but can be difficult to see in the dense thickets and tangled growth they prefer.

Where to see This antshrike is common in the lowlands and foothills of the southern Pacific humid forests. The male should not be confused with the male Black-crowned Antshrike as their ranges do not overlap. It is endemic to Costa Rica and Panama.

♀

♂

Black-crowned Antshrike *Thamnophilus atrinucha* 14cm

This species has a geographical range that includes Central America from Guatemala south, as well as parts of Colombia, Venezuela and Ecuador. Males in Costa Rica have black upperparts and grey underparts with bold white bars and spots on the wings. The female is brown and buff with buffy-white bars and spots on its wings. The Black-crowned Antshrike forages in forest thickets, old second growth and edge, taking katydids, stick insects and beetles as well as small lizards. It sometimes joins mixed-species flocks and occasionally attends swarms of army ants.

Where to see This is a common resident antshrike throughout the Caribbean lowlands and foothills, reaching as high as 700m in the north and 1,000m in the south-east.

♀

♂

Fasciated Antshrike *Cymbilaimus lineatus* 18cm

The male is a handsome antshrike, finely barred black and white, with a black crown. The female differs in being barred black and buff with a rufous crown. Both sexes have a heavy hooked bill and red eyes. The species forages in thickets and vine tangles, often along streams, feeding on diverse insects, spiders and occasionally small lizards.

Where to see It is a fairly common resident in most of the Caribbean lowlands, sometimes up to 1,200m. It is uncommon in the drier areas south of Lake Nicaragua.

♀

♂

Great Antshrike *Taraba major* 20cm

This is a bicoloured antshrike with a bushy crest, red eyes and a heavy bill. The male is black above and white below; the female rich rufous above and white below. This is a skulking bird of young second growth, dense scrub, tangled undergrowth and thickets of *Heliconia* and canebrakes along streams. It searches for insects, spiders and other arthropods but also takes any lizards and frogs that are encountered.

♂

Where to see This antshrike is a common resident of the lowlands and foothills, locally up to 1,000m on the Caribbean slope and as far north as Carara on the Pacific slope.

Dusky Antbird *Cercomacroides tyrannina* 14.5cm

The male is a slaty-grey to black bird with faint white barring on its wings and white tips to the tail feathers, while the female is brown above and rufous below. The slender bills of both sexes are tipped with a small hook. It forages in dense thickets and undergrowth at the forest edge, feeding on a great variety of insects and other invertebrates – crickets, caterpillars, wasps, beetles and spiders.

Where to see This is a common species in rainforest in the lowlands and foothills of both slopes, up to 1,000m on the Caribbean slope and occurring on the Pacific slope as far north as Carara.

♀

♂

Spotted Antbird *Hylophylax naevioides* 11.5cm

This is a small, short-tailed antbird that is very distinctive and easily identified. The male has a black throat, a chestnut back and rusty wing-bars, and a distinctive necklace of black spots on its white breast. The female is much duller but has the same broad rusty wing-bars. This is another antbird that is often seen accompanying army-ant swarms, feeding on fleeing insects and spiders.

Where to see This is a common species in rainforest in the lowlands and foothills of the Caribbean slope, up to about 800m and 1,000m in the south-east. It is also said to be abundant on the Pacific side of the Cordillera de Guanacaste between 600m and 750m.

Bicolored Antbird *Gymnopithys bicolor* 14.5cm

This is a very distinctive antbird, with chestnut-brown upperparts, white underparts, a blue patch around the eye and a shortish tail. The sexes are alike. This antbird is found low down in the forest understorey, often in small flocks and almost always foraging in association with a swarm of army ants, dropping down to seize the insects and spiders fleeing from the ants.

Where to see This species is a common resident in rainforest in the lowlands and foothills of the Caribbean slope and Pacific slope from Carara southwards. It occurs up to 1,500m on the Caribbean slope and 1,700m on the south Pacific slope.

Chestnut-backed Antbird *Poliocrania exsul* 14cm

The male is chestnut brown above with a black head and underparts. The female differs in being mostly chestnut brown with a brown-tinged, blackish head and neck, which differentiates it from the all-brown female Zeledon's Antbird. Both sexes have a conspicuous patch of bare blue skin around the eye. Chestnut-backed Antbirds skulk in dense thickets, foraging for insects, spiders and other invertebrates. The commonly heard song of two or three clear whistles is easily imitated.

Where to see This antbird is a common resident in rainforest in the lowlands and foothills of the Caribbean slope and on the south Pacific slope as far north as Carara, usually below 900m.

♂

♀

Zeledon's Antbird *Hafferia zeledoni* 19.5cm

The male is black except for its blue lores and patch of bluish skin behind the eye. The female is similar but dark brown instead of black. Both sexes have a broad tail that is regularly flicked downwards and raised slowly up. Zeledon's Antbird remains low in the forest understorey, foraging for diverse insects as well as occasional millipedes, scorpions, frogs and lizards. It also attends army-ant swarms but not as an obligate follower. Its song is a series of emphatic whistles: *teew, teew, teew, teew*.

Where to see It is found in forested foothills and low to mid-elevations the length of the country, up to 1,500m on the Caribbean slope and 900m on the Pacific slope.

Scaled Antpitta *Grallaria guatimalensis* 18cm

The Scaled Antpitta has a long-legged, tailless appearance. The scaly pattern on its upperparts can be hard to see but its grey cap and nape, and obvious pale moustachial stripe, are useful aids to identification. The sexes are similar. The Scaled Antpitta is often seen on trails at dawn. It forages on the forest floor, flicking leaves aside with its bill and probing soft ground for worms, snails, insects and small frogs.

Where to see It is an uncommon resident of middle elevations along the length of the Caribbean slope, mainly between 800m and 1,650m. It also occurs on the Pacific slope of the northern cordilleras above about 1,200m.

Streak-chested Antpitta *Hylopezus perspicillatus* 12.5cm

Formerly called Spectacled Antpitta, this small antpitta is a rotund, long-legged, short-tailed terrestrial bird with a buffy ring around its eyes ('spectacles'). It hops over the forest floor, flicking leaves aside with its bill to flush invertebrate prey, mainly insects and spiders. Its song is a series of mellow whistles, first increasing in tempo, then decreasing.

Where to see This antpitta is a fairly common resident in the lowlands of both the Caribbean and south Pacific slopes.

Ochre-breasted Antpitta *Grallaricula flavirostris* 10cm

This small species has a typical antpitta shape – plump, almost tailless and long-legged. The sexes are similar. Both have upperparts that are brown while the face, throat and breast are ochre, the latter lightly streaked darker. The best field marks are probably the buff eye-ring and yellow bill. In the field, the Ochre-breasted Antpitta can be shy and difficult to see well. It stays low, on or near the ground, gleaning insects, spiders and other invertebrates.

Where to see This species is scarce to uncommon, or perhaps overlooked due to its secretive habits and small size. It occurs in the wet, low to mid-elevations of the Caribbean slope, mostly between 750m and 1,300m. It is also recorded in a few areas on the Pacific slope of the Talamancas.

Yellow-bellied Elaenia *Elaenia flavogaster* 15cm

This species has greenish-olive upperparts, a white eye-ring and a rather bushy crest that, when raised, may expose a white crown-patch. It also has two buffy-white wing-bars and a yellowish belly. Its commonly heard call is a wheezy *breeer*. It is found in semi-open woodland, scrubby second growth and gardens. It catches small insects in flight but also takes many small fruits and arillate seeds.

Where to see This flycatcher is resident and common almost countrywide from the lowlands up to 1,850m. In many areas it has increased in numbers due to deforestation.

Mountain Elaenia *Elaenia frantzii* 15cm

Unobtrusive and easily overlooked, the Mountain Elaenia is rather drab overall. Unlike most elaenias, it has a rounded head that lacks a prominent crest. It also has an off white double wing-bar and white edging to the tertials. The Mountain Elaenia occurs in the canopy, on the forest edge and in overgrown clearings of highland forest. Like other elaenias, it sallies for insects and spiders, snatching them from vegetation, and also eats many berries and arillate seeds.

Where to see It is resident in the highlands of Costa Rica, above 1,200m in the northern cordilleras and above 1,850m in the Cordilleras Central and de Talamanca.

Mistletoe Tyrannulet *Zimmerius parvus* 9.5cm

This small, drab flycatcher has a small bill, slaty cap, white supercilium, pale eye, olive-green back and yellow edging to its wing feathers. It is rather inconspicuous in the field, where it forages in the middle and upper levels of forest, secondary growth and gardens, feeding on small insects, spiders and berries, especially mistletoe berries. It occasionally joins mixed-species flocks and its call is a clear, plaintive *peeeu*.

Where to see This flycatcher is a common resident countrywide, up to 3,000m, except for the north Pacific lowlands. It is relatively uncommon at higher altitudes.

Olive-streaked Flycatcher *Mionectes olivaceus* 12.5cm

Birds from Costa Rica and western Panama were formerly classed as a subspecies of the Olive-striped Flycatcher (which now has the scientific name *M. galbinus* and is found in eastern Panama and the Andes of South America). Good identification features for the Olive-streaked are the large pale spot behind its eyes and its streaky neck. It is often silent but the male does have an unusual song – a high-pitched, cicada-like trill that rises and falls. It is a small fruit-eating flycatcher, found in the lower or middle forest levels, that often gathers fruits by hover-gleaning. It often accompanies mixed-species flocks.

Where to see It is a common resident the length of both slopes at altitudes of 800m to 2,000m. After breeding, many birds move downslope to as low as 50m.

Ochre-bellied Flycatcher *Mionectes oleaginous* 12.5cm

This species has an olive-green head, upperparts and upper breast. The rest of its underparts are ochre and it has two buff wing-bars. The male is slightly larger than the female, but otherwise the sexes are similar. The Ochre-bellied Flycatcher has the habit of flicking one wing at a time up and over its back. It inhabits humid tropical forest and sometimes joins mixed-species flocks but also visits clearings and semi-open areas with fruiting bushes and trees. It takes some insects but feeds mostly on fruits and arillate seeds.

Where to see This species is a common to abundant resident in the lowlands and foothills of the Caribbean slope and the Pacific slope south of Carara, at altitudes up to 900m on the Caribbean slope and 1,200m on the Pacific slope.

Northern Scrub Flycatcher *Sublegatus arenarum* 14cm

This is a confusing flycatcher that is fairly small and nondescript and rather like a small elaenia. It has a short bill, obscure wing-bars and a clear, whistled call that is a good clue to its identity. It is almost restricted to mangroves but also occurs in adjacent scrubby woodland. It usually stays low, sallying and hovering to catch crickets, beetles, caterpillars and other insects. It also hawks insects in flight and takes some fruits.

Where to see This is a locally common resident around the Gulf of Nicoya and is rare in mangroves in the Golfo Dulce. It has not been recorded anywhere along the outer Pacific coast.

Common Tody-flycatcher *Todirostrum cinereum* 9.5cm

This is a tiny flycatcher with a large, flat bill, yellow underparts, a slaty-grey back and black wing feathers with yellow edges. It can be distinguished from its close relative, the Black-headed Tody-flycatcher *T. nigriceps* (not illustrated), by its conspicuous yellow eyes and grey back. The Common Tody-flycatcher is found in semi-open habitats, including gardens, with scattered trees and scrub, where it forages for caterpillars, beetles and other small insects. It often wags its tail from side to side.

Where to see This flycatcher is common throughout the country, except in forested areas that lack clearings, and at altitudes above 1,500m.

Eye-ringed Flatbill *Rhynchocyclus brevirostris* 15cm

This is a rather large-headed, greenish flycatcher with a short, flat bill and prominent white eye-rings. It is usually encountered inside mature forest, rarely at edges or in clearings. It is mainly insectivorous, taking katydids, caterpillars, beetles and other insects. It often accompanies mixed-species flocks, especially tanager flocks, and sometimes forages with army ants. It also eats some fruits and arillate seeds.

Where to see It is an uncommon to fairly common resident on the Caribbean slope and south Pacific slope north to Carara. It also occurs locally on the northern cordilleras up to at least 1,700m.

Yellow-olive Flatbill *Tolmomyias sulphurescens* 13cm

This species is relatively ordinary in appearance and quite similar to several other small flycatchers. It has a grey head with pale eyes and narrow, pale eye-rings that extend to the rather flat bill, giving it a 'spectacled' look. Its throat is pale grey, its back olive-green, its belly pale yellow, and its wing feathers have yellow margins. It occurs in both dry and humid wooded areas, including secondary growth, pastures and gardens. It feeds mainly on insects but also takes some berries.

Where to see This species is a common to abundant resident in the lowlands and foothills of the Pacific slope, up to around 1,000m, but is less common on the Caribbean slope, occurring up to 1,400m in the central highlands.

White-throated Spadebill *Platyrinchus mystaceus* 9.5cm

This is a tiny flycatcher with a large head and a distinctive facial pattern. It also has a usually concealed bright yellow crown, a very obviously broad and flat bill, and a short tail. The bold facial pattern includes a long yellowish supercilium, blackish ear coverts and a yellowish eye-ring. It forages in the understorey of wet cloud forest, where it typically sallies upwards to snatch small insects from the underside of leaves.

Where to see This spadebill is a common resident in wet rainforest at middle elevations the length of both slopes, mostly between 700m and 2,150m.

Northern Tufted Flycatcher *Mitrephanes phaeocercus* 12cm

This is a small flycatcher with a distinctive pointed crest. It is olive-green above with two buff wing-bars, and rich cinnamon below, becoming buffy-yellow on the lower belly. It is a bird of mature highland forests and second growth, foraging mainly in clearings and at edges, where it sallies from a perch to catch insects acrobatically, often returning to the same perch with a characteristic quiver of its tail. Its call is a cheery *chewee-chewee*. It is often seen in pairs throughout the year.

Where to see This flycatcher is resident in the highlands on both slopes, between 700m and 3,000m, but is most common in the middle of that range.

Yellowish Flycatcher *Empidonax flavescens* 12.5cm

This is one of the more brightly coloured *Empidonax* flycatchers. It is olive above and yellow-olive below, with a conspicuous, teardrop-shaped, pale eye-ring and two darker, yellowish wing-bars. This species has brighter plumage and a bigger, more conspicuous eye-ring than related species. It favours perches in the understorey, from which it sallies for small insects in typical flycatcher fashion, and also gleans insects from foliage and eats some berries.

Where to see This species is resident in cool highland cloud forest on both the Caribbean and Pacific slopes, mostly between 800m and 2,150m.

Black-capped Flycatcher *Empidonax atriceps* 11.5cm

This is a small dark flycatcher with a sooty-black head that sets off the prominent white eye-ring, which is incomplete but very broad behind the eye. It is a highland bird, found in the canopy of montane oak forest but also at much lower levels when on the forest edge, or in clearings and shady pastures. It catches flying insects, especially moths, flies and beetles, which are often taken in flight in short flights from an exposed perch.

Where to see This flycatcher is endemic to the highlands of Costa Rica and western Panama. It is a common resident of the higher parts of the Cordilleras Central and de Talamanca, mostly between 2,450m and 3,300m. During the rainy season some move to lower altitudes, down to 1,850m.

Social Flycatcher *Myiozetetes similis* 16cm

Except for its very small bill, the Social Flycatcher resembles a small Boat-billed Flycatcher or Great Kiskadee. Its conspicuous eye-stripe distinguishes it from the Grey-capped Flycatcher. It also has an orange to vermilion crown stripe, usually hidden, which is the reason that it is sometimes known as the Vermilion-crowned Flycatcher. The Social Flycatcher is found in agricultural areas, on the forest edge, in clearings with scattered trees and in shady gardens where it catches insects in flight or by snatching them from vegetation. It also takes small figs and many other fruits.

Where to see This flycatcher is resident and common in semi-open areas throughout the country, at elevations from the lowlands up to at least 1,700m.

Grey-capped Flycatcher *Myiozetetes granadensis* 16.5cm

This medium-sized species has a shorter bill than most other yellow-bellied flycatchers. Its back is olive-green and its wing and tail feathers are narrowly fringed with rufous. It resembles the Social Flycatcher, which shares much of the same range, but lacks the latter's bold, striped head pattern. As in other flycatchers, its vermilion crown-patch is usually concealed. It is usually found in shrubby second growth, on the forest edge and in gardens, where it sallies to catch insects in flight or hovers to take small berries.

Where to see This flycatcher is a common resident in the lowlands of the Caribbean and south Pacific slopes, locally up to 1,650m.

Great Kiskadee *Pitangus sulphuratus* 23cm

This is a large flycatcher that superficially resembles several other yellow-bellied flycatchers such as the Boat-billed Flycatcher. It is best distinguished from similar flycatchers by the bright rufous edging to the feathers in its wings and tail, and also by its distinctive strident call – *kis-ka-dee, kis-ka-dee*. Its varied diet includes small lizards, snakes and frogs, large insects, spiders, earthworms and fruit. It has also been seen diving into shallow pools to catch small fish and tadpoles.

Where to see This is a bird of open country with scattered trees. Thanks to deforestation, it is now an abundant resident throughout most of the country from the lowlands to 1,500m.

White-ringed Flycatcher *Conopias albovittatus* 16cm

This is one of several similar medium-sized flycatchers that are bright yellow below with bold black and white head patterns. Among these flycatchers, the White-ringed has a noticeably thin bill and a broad white supercilium that wraps around the back of the head (hence its name). It forages in trees in the semi-open, on the forest edge or in trees bordering rivers or lakes. There, it scans foliage below, before darting down to catch insects and spiders. It also gathers berries while in flight.

Where to see The White-ringed Flycatcher is a locally common resident in the Caribbean lowlands, occasionally encountered as high as 600m.

Golden-bellied Flycatcher *Myiodynastes hemichrysus* 20cm

This species is yellow below, dark olive above and has a white stripe both above and below its eye, a black stripe through the eye and a grey moustachial stripe. This head pattern distinguishes it from other flycatchers with yellow bellies. The Golden-bellied Flycatcher occurs in pairs and family parties that are conspicuous because of their noisy habits. The common call is a squeaky *seeeik* and the dawn song is a melodious repeated *tre-le-loo*. Golden-bellied Flycatchers typically hunt from fairly high perches, hawking flying insects or snatching exposed prey, or sometimes berries, from foliage.

Where to see It is a fairly common resident in wet montane forest, found at 700m to 1,850m along the whole length of the Caribbean slope. It is less widespread on the northern cordilleras along the Pacific slope.

Sulphur-bellied Flycatcher *Myiodynastes luteiventris* 20cm

This is a large, streaky flycatcher with a bold white supercilium and moustache, a yellow belly and a bright rufous tail. It also has a concealed golden-yellow patch on its crown. It is a noisy bird and its shrill calls sound like rusty hinges or squeaky toys. The Sulphur-bellied Flycatcher catches insects in flight, including katydids, caterpillars and beetles, but also eats at least some berries.

Where to see This species is mostly seen in dry forest or on the edges of wetter forest. It is a common passage migrant on both slopes from the lowlands up to middle elevations. It is also a breeding resident on the northern half of both slopes, from the lowlands to 2,000m on the Pacific slope but not below 2,000m on the Caribbean side. It arrives by March or early April and departs by mid-October.

Streaked Flycatcher *Myiodynastes maculatus* 20cm

This species resembles the Sulphur-bellied Flycatcher, but has a heavier bill with a pale pinkish base to the lower mandible, a weaker moustachial stripe and little or no yellow on its underparts; it also has a clucking rather than squeaky call. It has a similar rufous rump and tail and an often concealed golden-yellow crown-patch. It hawks for flying insects, especially wasps and ants, but also hovers close to vegetation, searching for small lizards and insects. It also takes many small fruits.

Where to see This flycatcher is resident and a passage migrant on the north Pacific slope, common in the lowlands but also occurring up to 1,000m.

Boat-billed Flycatcher *Megarhynchus pitangua* 23cm

This is a very large, handsome flycatcher with a big, heavy bill. It has a dark head with a conspicuous white supercilium, a white throat and a usually concealed orange crown-patch. It has dull olive upperparts and yellow underparts. It is more olive above than the similar Great Kiskadee and lacks the latter's rufous wing-panels. It also has a larger bill. The Boat-billed Flycatcher favours the forest edge, including the canopy, particularly along rivers. It eats large insects, including many cicadas, and takes fruit by both sallying and gleaning.

Where to see It occurs throughout Costa Rica up to 1,850m or higher, mainly in semi-open areas with taller trees. Some descend to lower altitudes when not breeding.

Tropical Kingbird *Tyrannus melancholicus* 21cm

This species has a grey-olive back and yellow underparts, tinged with grey on the breast. It has an all-dark notched tail, large bill and a vermilion patch on the crown that is usually concealed, except when the bird is excited. Males and females are alike. They perch in open places, ready to pursue insect prey with what can be impressive aerial acrobatics. They are often seen perched on utility wires and fences.

Where to see It is common throughout the country up to altitudes of 1,850m, occasionally higher. The similar Western Kingbird *T. verticalis* (not illustrated) is a sporadic migrant visitor to the Pacific slope, mainly in Guanacaste. It can be distinguished by its less distinct dark mask and white outer tail feathers.

Scissor-tailed Flycatcher *Tyrannus forficatus*

19.5cm (plus 15cm for male's tail, 9cm for female's)

Males of this elegant, long-tailed flycatcher are silver-grey above with darker wings and white underparts. Much of the plumage is tinged with pink, and the pink underwing linings are visible in flight. Females are similar to males but duller with a shorter tail. This flycatcher is usually encountered in open savannah country where it often perches on fences or barbed wire, waiting for its insect prey. It also eats a few berries. In the evening, hundreds converge at communal roosts.

Where to see This flycatcher is a locally common to abundant winter visitor to the dry north-west lowlands, usually arriving in late October and departing by mid-April. Also occurs rarely in the Valle Central and on the south Pacific slope.

Rufous Mourner *Rhytipterna holerythra* 20cm

This species has entirely rufous plumage, though it is slightly paler on its underparts. It can be confused with the Rufous Piha *Lipaugus unirufus* (not illustrated), which is bigger and has a very different voice (often a loud, explosive *pee-haa*). The Rufous Mourner's call, on the other hand, is a slow, mournful wolf-whistle: *whee-wheeu*. When foraging, the Rufous Mourner sits still for long periods, scanning foliage, waiting to sally out and catch prey such as stick insects, katydids, caterpillars and beetles. It also snatches berries while in flight.

Where to see This is a common resident in wet forest, adjacent secondary forest and shady plantations in the lowlands and foothills of the Caribbean and south Pacific slopes, up to altitudes of about 1,200m.

Dusky-capped Flycatcher *Myiarchus tuberculifer* 16.5cm

This is a medium-sized flycatcher that is quite small for a *Myiarchus*. It is extremely similar in appearance to other *Myiarchus* flycatchers and most easily identified by its contrasting blackish cap, fairly bright yellow belly and its call – a plaintive, whistled *peeur* or *wheeeeeur*. The Dusky-capped Flycatcher inhabits wooded areas with clearings and gaps, and has the typical flycatcher habit of catching its insect prey in flight. It also eats many fruits, including berries and arillate seeds.

Where to see It is a common resident throughout the country from the lowlands up to about 1,200m and occasionally as high as 1,850m.

Brown-crested Flycatcher *Myiarchus tyrannulus* 19cm

This is a fairly large flycatcher with a heavyish, hooked bill. The bill is all black, which is helpful in distinguishing this species from the very similar Great-crested Flycatcher, which has a pale base to the lower mandible as well as being a little more brightly coloured. It can also be distinguished by its call, transcribed as a loud 'come here, come here' or 'whit-will-do, whit-will-do'. When disturbed, it has a habit of bobbing its head like a lizard. Brown-crested Flycatchers forage in open country with scattered trees and scrub, feeding on insects and berries.

Where to see Inhabits woodland habitats in the lowlands and foothills of the northern Pacific slope, south to Orotina, from sea level up to 900m.

Bright-rumped Attila *Attila spadiceus* 18cm

The Bright-rumped Attila appears to be big headed, with a large bill that is hooked at the tip. Its plumage is variable in colour: its upperparts range from olive-green to brown, with varying amounts of black streaking, and its breast can be yellow, olive or grey, often streaked with brown. It usually has a yellow rump. It has a very upright stance and constantly wags its tail up and down. Its voice is loud, assertive and easily identified once heard. It is an active, noisy bird, foraging in search of its varied diet that includes small lizards, frogs, insects and spiders as well as berries and arillate seeds.

Where to see The Bright-rumped Attila is resident and common wherever patchy forest remains, from the lowlands up to at least 1,850m.

Bare-necked Umbrellabird *Cephalopterus glabricollis* ♂41cm ♀36cm

The male is all black with a large umbrella-like breast, plus bare red skin on its throat and chest that forms an inflatable sac and wattle. The female is a little smaller and duller black with a smaller crest. At dawn, males display in the canopy, making a double hoot call. Umbrellabirds are usually seen in primary forest where they feed on large fruits and insects, small lizards and frogs.

♀

Where to see This species is endemic to the Talamancan montane forests of Costa Rica and Panama. It spends most of the year in the lowlands and foothills of the Caribbean slope, usually below 500m. It moves upslope to breed at higher elevations (800–2,000m).

♂

Three-wattled Bellbird *Procnias tricarunculatus* 25–30cm

The spectacular male has a huge black gape, three worm-like wattles and immaculate chestnut and white plumage. The female is very different, with olive-green upperparts and yellowish underparts streaked, with dark olive-green. The male is famous for its loud, far-carrying, bell-like call. Bellbirds are frugivorous, feeding on wild avocados and other fruits.

Where to see It breeds in highland forest, mostly at altitudes of 1,200–2,300m. Outside the breeding season, it descends to the foothills and lowlands of both slopes and wanders widely, depending on the availability of fruit.

♂

♂ ♀

Snowy Cotinga *Carpodectes nitidus* 22.5cm

Both sexes have broad wings, a rather short tail and a high dove-like forehead. The male is basically white. The female differs in being much greyer, especially on the head, and its secondaries and wing-coverts are edged white. Small groups frequent the canopy and semi-open forest, feeding on wild avocados, figs and many other fruits.

Where to see This cotinga is a scarce resident in forested areas throughout the Caribbean lowlands up to about 750m. A good place to see it in the non-breeding season is at La Selva Biological Station.

♀

♂

White-ruffed Manakin *Corapipo altera* 10cm

Males of this manakin are glossy blue-black with a very conspicuous erectile white throat-patch. Females are mostly olive-green, paler below, with a grey throat. The displays of male White-ruffed Manakins, centred on mossy logs, consist of slow, fluttering 'butterfly' flights to and from their logs and nearby perches. Both sexes sometimes join mixed-species flocks, often sallying upwards to snatch berries or insects.

Where to see This manakin is common on the Caribbean slope, breeding mainly between 400m and 900m, and between 1,100m and 1,500m on the south Pacific slope. It is an altitudinal migrant that descends to the foothills and lowlands in the non-breeding season.

♂

Long-tailed Manakin *Chiroxiphia linearis*

11.5cm (plus 10–15cm for male's tail, 2–3cm for female's)

There are several other manakins that are similar to the Long-tailed Manakin but none have its very long tail and none overlap its range. Male Long-tailed Manakins, usually two or three together, perform courtship displays to attract females in dense undergrowth. This manakin typically frequents dry forest, gallery forest and tall secondary growth where it forages for a considerable diversity of small fruits and a few insects.

Where to see This spectacular manakin is a locally common to abundant resident in the lowlands and foothills, up to 500m, on the Pacific slope south to Carara and the Valle Central. Two good places to see the Long-tailed Manakin, including displaying males, are Monteverde and Villa Lapas.

Velvety Manakin *Lepidothrix velutina* 9cm

The male is velvety black except for its bright blue cap. The female is a rather bright green, less olive than most other female manakins, and paler below. The males' courtship display is relatively simple, consisting of darting flights to supplant other males, accompanied by song but without wing-snaps. It forages for both small fruits and insects.

Where to see This manakin is a common resident of rainforest and second growth on the south Pacific slope as far north as Carara National Park, from the lowlands up to 1,350m. It also occurs in the extreme south-east of the Caribbean lowlands.

White-collared Manakin *Manacus candei* 11cm

The White-collared Manakin is another small, plump species. The unmistakable males have a black crown, a black band across their back, an olive-green rump and yellow belly. Females are olive-green with a yellow belly and orange legs, and are very similar to female Orange-collared Manakins, but there is no range overlap. When breeding, males are involved in lekking behaviour on the forest floor. Each male clears a patch of bare earth, and leaps to and fro between thin saplings, giving a loud wing-snap. Away from display areas, these manakins are usually found in ones and twos, feeding at fruiting trees and bushes, mainly in the forest understorey and often in *Heliconia* thickets.

Where to see This manakin is a common resident in humid forest edge and second growth in the lowlands and foothills of the Caribbean slope, from sea level to 700m.

♂

♂

♀

Orange-collared Manakin *Manacus aurantiacus* 10cm

This is a small, plump manakin. The adult male has a black crown, wings and tail, and a black band across the middle of its back. The rest of its head, neck, breast and upper back are orange, the rump is olive-green, and the belly is bright yellow. Females and young males are olive-green with a yellow belly. They are very similar to the female White-collared Manakin, but there is no range overlap. Like other manakins, this species has a fascinating breeding display at a communal lek. Each male clears a small patch of forest floor up to 120cm across, and leaps to and fro between thin upright bare sticks, giving a loud wing-snap. When a female is present, males jump together, crossing each other above the bare display court. This species is found in the forest understorey, typically at a male display site or a fruiting tree. It feeds low in trees and shrubs on fruit and a few insects.

Where to see It is found in the lowlands and foothills of the Pacific slope, up to 1,100m, and is replaced on the Caribbean slope by the closely related White-collared Manakin. It is endemic to Costa Rica and Panama.

Red-capped Manakin *Ceratopipra mentalis* 10cm

The male has a scarlet head, white eyes, a black body and yellow thighs. As with most other manakins, the female is dull olive green. When breeding, males gather in a loose, well-dispersed assembly and display, often including wing-snapping, to attract females. Like other manakins, this species is mainly frugivorous, foraging for a variety of small fruits and a few insects.

Where to see This species inhabits the lower and middle levels of mature rainforest and old second growth. It is a common resident in the lowlands and foothills of the Caribbean and south Pacific slopes, locally up to 1,050m.

Sulphur-rumped Myiobius *Myiobius sulphureipygius* 12cm

This is an attractive, small flycatcher that always appears very active, lowering its wings to display its pale-yellow rump and fanning its ample tail. It is often found close to forest streams, where it forages low down, usually at eye level, sallying for insects of many different kinds. It often joins the mixed-species foraging flocks that roam through the lower levels of the forest and sometimes attends swarms of army ants.

Where to see This species is a common resident on the Caribbean slope and Pacific slope south of Carara. It occurs from sea level up to about 800m, higher in the south-east.

Masked Tityra *Tityra semifasciata* 21cm

The male is a striking bird, mainly white with a black mask, wings and tail. The female differs in having a dark brown mask and cap, and being tinged greyish brown. Both sexes have bare red skin at the base of the bill and around the eyes. Their calls are dry grunting croaks. They feed on many fruits, including figs and small wild avocados, as well as large insects.

Where to see This tityra is a common resident in forest and semi-open habitats in the lowlands and middle elevations up to 1,500m on both slopes, sometimes higher.

Barred Becard *Pachyrhamphus versicolor* 12cm

This becard is easily identified because it is the smallest Costa Rican becard and the only one with barred underparts and a prominent eye-ring. Males have a yellow face, black cap and back, and thin barring on the underparts. Females have a grey cap, rufous wings, and yellow underparts with only faint barring. The Barred Becard is found in the middle and upper levels of cloud forest and on the forest edge. It often accompanies mixed flocks, sallying for insects and spiders, and sometimes hovering to take berries.

Where to see It is an uncommon resident in cloud forest at altitudes of 1,500–2,500m from the Cordillera de Tilarán south on both slopes to the Cordillera de Talamanca.

Cinnamon Becard *Pachyramphus cinnamomeus* 14cm

This is a mainly uniform rufous bird with a pale stripe above the eye and a pale throat and underparts. Unusually for becards, the sexes are similar. They superficially resemble the Rufous Mourner and Rufous Piha *Lipaugus unirufus* (not illustrated), but are smaller and less uniform in colour. The Cinnamon Becard occurs in relatively open habitats, including the forest edge, stream margins and in semi-open woodland. It is active when foraging, typically making short flights to snatch insects and fruits. It has a constantly uttered, plaintive, piping call.

Where to see It is a common resident on the Caribbean slope up to 750m in the foothills. It is uncommon on the Pacific slope from the Panama border north to the Gulf of Nicoya, usually in or close to mangroves.

White-winged Becard *Pachyramphus polychopterus* 14.5cm

♂

♀

This becard has a relatively large head and thick bill. The male is black and grey with a dark cap and white markings on the wing. The female is olive-green above, buff below, with buff markings on the wings, a pale throat and a pale eye-ring. The dark cap often looks less sleek than in the photographs. It favours lightly wooded areas where it forages at middle to high levels for small invertebrates and fruits.

Where to see It can be found in the lowlands of both slopes but is least numerous in dry areas.

Black-and-white Becard *Pachyrhamphus albogriseus* 14cm

The male Black-and-white Becard is mostly grey with bold white bars and edgings on its wings, a black cap and pale lores. The female is very different, with a rufous cap bordered with black, a thin white eye-ring, pale yellow belly and olive back. It is mostly seen quite high in the forest, just below and in the canopy. Pairs often accompany mixed-species flocks, foraging and often hovering to snatch insects and fruit.

Where to see This becard is an uncommon resident of mid-level rainforest, between 800m and 1,850m, from the Cordillera de Tilarán south to Panama.

♂

Rose-throated Becard *Pachyrhamphus aglaiae* 16.5cm

This species is relatively large, big headed and heavy billed. In Costa Rica, unlike in the drier, more northern parts of its range, the pink throat-patch it is named for is faint or absent. The male is slaty grey, darker on the back and crown. The female has a blackish crown but is otherwise tawny-rufous above and buffy below. The Rose-throated Becard is found in deciduous woodland, gallery forest and scrub. It takes many different insects, spiders and other invertebrates as well as many fruits.

Where to see This species is a fairly common resident in the lowlands and foothills, up to 300m, of the north Pacific slope south to about Orotina.

♀

Red-eyed Vireo *Vireo olivaceus* 14cm

This vireo is plain olive-green above and clean white below with a strong head pattern and no wing-bars. The red eye of the adult is not always easy to see, unlike the white supercilium, which is bordered above and below by blackish lines. The flanks and undertail have a greenish-yellow wash. Red-eyed Vireos usually forage in trees or large shrubs but also venture into parks and gardens. They feed on the usual insects and spiders but also take many small berries and arillate seeds.

Where to see This is an abundant passage migrant in both autumn and spring. It is most commonly seen in the lowlands but it can be seen up to 1,500m, and some reach as high as 2,150m.

Mangrove Vireo *Vireo pallens* 11.5cm

In Costa Rica, though not elsewhere in its range, the Mangrove Vireo is found only in or very close to mangroves. It is rather dull-coloured but has a yellowish stripe between the pale eye and the bill, and two white wing-bars. It seems to be most easily encountered in red mangroves (*Rhizophora*). The Mangrove Vireo usually stays fairly low, foraging for small insects and spiders. It also takes some small fruits.

Where to see The Mangrove Vireo is locally resident in mangroves on the northern Pacific coast from the Nicoya Peninsula to as far south as Tivives.

Azure-hooded Jay *Cyanolyca cucullata* 29cm

This is a medium-sized, mostly blue jay with a black face and throat, topped by a sky-blue patch on its crown and nape. It forages in pairs or small flocks and often joins mixed-species flocks that forage through the middle levels of humid cloud forest, probing actively for food amongst foliage, epiphytes and bark. The Azure-hooded Jay is omnivorous, eating small frogs and lizards, katydids, beetles and other insects, spiders, berries and seeds.

Where to see This species is widespread but not numerous in cloud forest at 800m to 2,100m on the Caribbean slope from the Cordillera de Tilarán south to Panama. It is also quite common on the Pacific slope of the Cordillera de Tilarán.

Brown Jay *Cyanocorax morio* 39cm

This is a large jay with a graduated, white-tipped tail. Its head, upperparts and breast are dark brown, gradually shading to creamy-white on its belly. The sexes are alike. Juveniles are similar to adults but have a yellow bill, legs and eye-ring. It is an extremely social, raucous bird, usually seen in small, noisy groups that call *pyaaaah-pyaaaah*. It is an opportunistic feeder, taking more or less anything edible that it can find, including eggs and nestling birds, lizards, frogs, insects, spiders, varied fruits and nectar.

Where to see This jay is a common resident of deforested parts of Costa Rica from sea level to 2,500m. Its range continues to expand where deforestation continues.

White-throated Magpie-jay *Calocitta formosa* 46cm

With its blue and white plumage, long tail and curly crest, the White-throated Magpie-jay is an unmistakable bird. Typically, it travels in flocks of 5–10 birds, foraging for small frogs and lizards, birds' eggs and nestlings, and large insects. It also takes many fruits, both wild and cultivated, and sips nectar from large flowers.

Where to see This species is a common, noisy resident on the north Pacific slope up to about 800m. It is found in scrub, savannah and gallery forest, and along wooded water courses; also in gardens.

Long-tailed Silky-flycatcher *Ptilogonys caudatus* ♂24cm ♀21cm

This attractive grey-blue and yellow bird has a distinctive pointed crest and long black tail. The female is similar, but duller and more olive. Except when breeding, this species is usually seen in small straggling flocks, often perched on exposed branches of tall trees. It feeds on insects that are caught in flight after long aerobatic chases. It also takes fruit, particularly mistletoes.

Where to see This silky-flycatcher is a common resident in the Cordilleras Central and de Talamanca, mainly from 1,850m up to the timberline. Some wander as low as 1,200m outside the breeding season.

Black-and-yellow Phainoptila *Phainoptila melanoxantha* 21cm

The male of this species is black with a yellow rump, breast and flanks, and a grey belly. The female is less distinctive, with black colouring confined to the cap; the face is grey, back olive-green, and the underparts are mainly olive-grey, with yellow flanks. This silky-flycatcher inhabits highland forests and clearings, feeding mainly on berries but also insects, which it gleans from foliage or catches while in flight.

Where to see This species is a fairly common resident on mountains throughout the country, above 1,200m on the northern cordilleras and above 1,850m on the Cordilleras Central and de Talamanca. It is endemic to Costa Rica and Panama.

♀

♂

Mangrove Swallow *Tachycineta albilinea* 13cm

This is a small swallow, metallic blue-green above and white below, with a large white rump patch. It also has a distinctive white line above its black lores. The Mangrove Swallow forages for insects over open water and so frequents coastal mangroves and estuaries, as well as lakes or rivers well inland. It subsists primarily on a diet of small flying insects but also takes some larger prey, including dragonflies and bees.

Where to see The Mangrove Swallow is an uncommon to locally abundant resident on both slopes, mainly in the lowlands but up to 1,000m if suitable habitat is available.

Blue-and-white Swallow *Pygochelidon cyanoleuca* 11cm

As its name implies, this small swallow is blue and white, with the blue darkening to black on the wings and tail. It has a short, notched tail and relatively short wings. In flight it is most easily distinguished from other small swallows by its dark undertail-coverts, which contrast with its white belly. It is a swallow of open areas, including towns and villages, farms, and forest clearings. It often perches on overhead wires. Like other swallows, this species hawks for insects on the wing and is attracted to termite swarms.

Where to see This swallow is common and resident countrywide from 400m to 3,000m or higher. It is rare in Guanacaste.

Southern Rough-winged Swallow *Stelgidopteryx ruficollis* 12cm

This is a plain, brownish swallow. It is very similar to the Northern Rough-winged Swallow *S. serripennis* (not illustrated), but look for this species' contrasting pale rump in flight and tawny throat. It is found in open habitats, especially close to lakes, streams and roads, hawking for small insects, including flies, flying ants, wasps and beetles, and often gathers in small loose flocks.

Where to see This is a common to abundant swallow throughout the lowlands and foothills of the Caribbean and south Pacific slopes.

Band-backed Wren *Campylorhynchus zonatus* 16.5cm

This is a readily identified large wren, boldly barred black and white above, and spotted below, with a cinnamon belly. It usually occurs in small noisy flocks or extended family groups of 4–10 birds that stay in contact with their grating calls. Members of a group forage at middle to high levels in forest but sometimes enter forest clearings or even gardens. They clamber over branches, probing bromeliads and crevices in bark, searching for insects and spiders. Sometimes they join mixed-species flocks.

Where to see This wren is a common resident in forested areas on the Caribbean slope from the lowlands up to about 1,700m.

Rufous-backed Wren *Campylorhynchus capistratus* 17cm

This is a large wren – white below, barred above, with a striking black cap and white supercilium, and a relatively large, downcurved bill. It is insectivorous and forages in small family groups in dry, deciduous woodland, secondary growth, scrub and gardens.

Where to see This wren occurs in the lowlands and foothills of the north Pacific slope from sea level up to 800m.

Riverside Wren *Cantorchilus semibadius* 13cm

This is a medium-sized wren with rufous upperparts, bold black-and-white barred underparts and a complex black-and-white face pattern. The sexes are alike. The Riverside Wren has varied calls and songs that are often sung as duets by a pair. It inhabits forest edge, dense thickets alongside watercourses, and the swampy edges of ponds. Like many wrens, this species probes and gleans actively in dense thickets and undergrowth, foraging for insects and other invertebrates. It sometimes joins in with other species foraging at army-ant swarms.

Where to see The Riverside Wren is found along the Pacific slope from the Gulf of Nicoya to western Panama. It ranges from sea level to 1,200m.

Bay Wren *Cantorchilus nigricapillus* 14.5cm

This is a rufous-chestnut wren with a striking black-and-white head pattern and a white throat. It favours dense thickets often bordering streams. Its loud distinctive song consists of rich-toned whistles, warbles and trills. It forages in thickets and vine tangles for small invertebrates, including crickets, beetles, caterpillars and spiders.

Where to see This is a common resident of forested areas throughout the length of the Caribbean slope, from sea level to about 700m in the north and 1,000m in the south.

Stripe-breasted Wren *Cantorchilus thoracicus* 11.5cm

This is a medium-sized wren, brown above with bold black-and-white streaking on its face, throat and breast. The streaking is distinctive within its range – there are other wrens that are barred or spotted but not streaked. The Stripe-breasted Wren is often found in vine tangles and dense vegetation in clearings and alongside streams. When foraging, it mostly stays fairly low, poking into dead leaves and moss as it hunts for insects and spiders.

Where to see This species is a common resident of the lowlands and foothills of the Caribbean slope from sea level up to 1,000m. It also reaches the Pacific slope through low passes in the northern cordilleras.

House Wren *Troglodytes aedon* 10cm

This is a small brown wren, lacking any strong markings, though the wings and upper tail have barring, and the belly and supercilium are a pale buff. It is a common species throughout most of the country, particularly around habitation, and often roosts and nests in crevices in buildings. It is smaller and has less clearly defined markings than the Plain Wren *Cantorchilus modestus*, and has paler, less rich colouring than the smaller Ochraceous Wren *Troglodytes ochraceus* (neither illustrated).

Where to see It can be found more or less throughout the country except at high elevations above 2,750m. It is rare only in extensively forested regions and in the dry, Guanacaste lowlands.

Timberline Wren *Thryorchilus browni* 10cm

This is a very small, stub-tailed wren with that is rufous above and dirty greyish below with a white supercilium. It is quite similar to the wood wrens but has white markings on its wing feathers. It is mostly found in bamboo thickets or the edge of oak forest, where it searches for small insects, caterpillars and spiders. Its high-pitched song is a complex series of whistles and trills that sometimes give the impression of a squeaky swing.

Where to see This wren is a common resident at high altitude, mostly above the timberline, from the Cordillera Central south along the Cordillera de Talamanca. It is endemic to Costa Rica and Panama.

White-breasted Wood Wren *Henicorhina leucosticta* 10cm

This is a very small, stub-tailed wren that is common in lowland rainforest, usually on or near the ground in the forest understorey. Note its boldly striped face, white bib, rich dark brown upperparts and short tail. It is heard far more often than seen and is notably elusive, even in fairly open understorey. The common call is a metallic *peenk*, which is rather frog-like and difficult to trace. It feeds on small invertebrates, including katydids, caterpillars and spiders.

Where to see This wood wren is a common resident in the lowlands and foothills of the Caribbean slope. It also reaches the Pacific slope of the northern cordilleras. On the south Pacific slope it is common in the foothills at altitudes of 300–1,850m as far north as Carara.

Grey-breasted Wood Wren *Henicorhina leucophrys* 11cm

This is another small wren with a short, frequently cocked tail. It has brown upperparts, a grey breast, a dark grey cap and streaking on its face, and a white supercilium. It has a rich whistled song with a rollicking cadence, distinct from songs of other wrens in the same habitat. The Grey-breasted Wood Wren forages in low, dense vegetation, busily searching for insects and other invertebrates in clusters of leaves caught up in trees, leaf litter or other debris.

Where to see This is a common resident wren of humid highland and foothill forest on both slopes, from 800m to the timberline on the Caribbean slope and from 1,100m to the timberline on the Pacific slope.

Trilling Gnatwren *Ramphocaenus melanurus* 12cm

This species was formerly called the Long-billed Gnatwren. The long, slender bill with a small terminal hook is its most distinctive feature. It also has a longish, graduated, white-tipped tail.

It has a skulking habit so can be difficult to see in the dense, tangled forest edges that it inhabits, though its song, a prolonged trill, often attracts attention. It forages in a wren-like manner, searching for small insects and spiders.

Where to see The Trilling Gnatwren is a common resident countrywide from the lowlands up to 1,200m. It is less numerous in the drier north-west except in evergreen vegetation close to water.

Tawny-faced Gnatwren *Microbates cinereiventris* 10cm

This gnatwren's bill is shorter than that of the Trilling, but its tawny face and white throat make it easy to identify. In some ways it resembles a wren with its short and often cocked tail. The sexes are alike. The Tawny-faced Gnatwren forages through leaf litter and low vegetation, searching for small insects and spiders. It regularly joins mixed-species foraging flocks but rarely follows army-ant swarms.

Where to see This gnatwren is a fairly common resident on the Caribbean slope from the lowlands up to 900m in the north and 1,200m in the south-east.

Black-faced Solitaire *Myadestes melanops* 17cm

This is a dark, slate-grey bird with a black mask, wings and tail. The bill and short legs are orange. It is celebrated for its exquisite song of high, fluty whistles which has made it a target for the cage-bird trade. It forages at all levels, from the understorey to the canopy, eating a lot of fruit but also insects, spiders and earthworms.

Where to see When breeding, this bird is a common resident of wet mountain forest at altitudes between 900m and 2,750m on both slopes. After breeding, many birds move to lower altitudes, often to 450m or lower. It is endemic to Costa Rica and Panama.

Wood Thrush *Hylocichla mustelina* 18.5cm

The Wood Thrush is a boldly patterned thrush that has warm rusty-brown upperparts and white underparts with bold black spots. It winters in forest undergrowth, secondary thickets and overgrown banana and cacao plantations. It hops on the ground, searching and probing the leaf litter for insects, spiders and earthworms, and also eats a lot of small fruits.

Where to see This thrush is an uncommon fall passage migrant, from late September to mid-November, and spring migrant in March and April, mainly along the Caribbean coast. In between, it is a widely distributed but uncommon winter resident on both slopes, from the lowlands up to 1,400m and often higher.

Black-headed Nightingale-thrush *Catharus mexicanus* 16cm

This thrush is a small, plain-breasted species found in the understorey of primary rainforest. It is easily identified by its blackish cap, contrasting orange eye-ring and bill, and dark brownish-olive upperparts. Its song, which includes flute-like phrases, trills and whistles, is rather scratchy but not unattractive. The Black-headed Nightingale-Thrush feeds mainly on or close to the ground, searching leaf litter and moss for katydids, caterpillars, beetles and other spiders. It also takes a lot of small fruits.

Where to see This species is a common resident of foothill forest on the Caribbean slope at altitudes of 300m to 800m. It also occurs in rather drier forest on the Pacific slope of the northern cordilleras at 700m to 1,300m.

Slaty-backed Nightingale-thrush *Catharus fuscator* 17cm

The Slaty-backed Nightingale-thrush has a dark grey back and head, a medium grey throat and a light grey belly. This rather dull plumage contrasts with its bright orange legs and bill. It also has a pale grey iris which is distinctive amongst similar nightingale-thrushes. The sexes are alike. Males have a simple flute-like song of alternating phrases. This species almost always forages close to the ground, taking the usual mix of insects and small fruits.

Where to see This small thrush is a common resident at middle elevations, from 800m to 2,300m, on the Caribbean slope and locally on the Pacific slope, mainly on the Cordillera de Tilarán.

Swainson's Thrush *Catharus ustulatus* 16cm

Swainson's Thrush is easily distinguished from other migrant thrushes by its very distinctive buffy eye-ring. Also, its spots are clearly defined on the upper breast but poorly defined and smudgy towards the belly. Swainson's Thrush is typically encountered in the lower levels of rainforest and old secondary growth, where it feeds mainly on varied fruits and arillate seeds. It also takes some insects and occasionally follows army-ant swarms. Swainson's Thrush is often heard singing while on spring migration. Its attractive song is a rather flute-like series of upwardly spiralling trills.

Where to see Swainson's Thrush is an abundant passage migrant on both slopes in both fall and spring. It is also a winter visitor on both slopes from sea level to about 1,500m.

Black-billed Nightingale-thrush *Catharus gracilirostris* 14.5cm

This thrush is small and skulking, and is more often heard than seen. It is distinctive if seen well, differing from related species in having a dark eye, a black bill and brownish legs. The sexes are alike. The Black-billed Nightingale-Thrush normally forages on the forest floor, hopping and making frequent stops. It turns leaf litter in typical thrush fashion, seeking insects and spiders. It also eats many small fruits. This species is often tame and confiding.

Where to see This nightingale-thrush is endemic to the highlands of Costa Rica and western Panama. It is common to abundant in oak forest and second growth in the Cordilleras Central and de Talamanca from about 2,000m to above the timberline. It sometimes occurs in páramo up to at least 3,500m.

Sooty Thrush *Turdus nigrescens* 25.5cm

The Sooty Thrush is a large thrush, endemic to the highlands of Costa Rica and western Panama. It is all black with pale grey eyes and a yellow eye-ring, bill and legs. The female is browner than the male. The Sooty Thrush is found singly or in pairs at high elevations, usually above 2,200m, often in the open, hopping around on the ground but or at the forest edge, where it searches for insects and also berries, especially Ericaceae, *Solanum*, melastomes, blackberries and arillate seeds.

Where to see This thrush is a common to abundant resident of high elevations in the Cordillera Central and Cordillera de Talamanca.

Mountain Thrush *Turdus plebejus* 24cm

This is a rather dull-coloured thrush that resembles other *Turdus* thrushes in general appearance and habits. Adults are uniformly dull olive-brown with faint white streaks on the throat. The bill is black and the legs are dark brown. The male's song is most untypical for a thrush – a mechanical succession of weak, unmusical notes with little variation in pitch. The Mountain Thrush is a bird of epiphyte-laden mountain forest, adjacent clearings and farmland with scattered trees. It feeds in fruiting trees and bushes, less often on the ground, turning leaf litter in search of insects, spiders, snails and earthworms.

Where to see It breeds in the mountains throughout from 1,300m up to the timberline. Following the breeding season, it forms flocks and descends to lower altitudes of 900m or less.

White-throated Thrush *Turdus assimilis* 23cm

The White-throated Thrush has a distinctive black-and-white streaked throat, bordered below by a white crescent. Its back is dark greyish olive, its breast is light grey, and it has a bold yellow eye-ring, bill and feet. It is a superb songster. It is a widespread thrush of forest and woodland in the foothills and highlands, though some wander to the lowlands in winter. It forages for insects, earthworms and also many fruits.

Where to see It breeds in the foothills and lower mountain slopes, mainly on the Caribbean side, at 800m to 1,850m. After breeding, it descends to lower elevations.

Pale-vented Thrush *Turdus obsoletus* 23.5cm

This is a large, rather drab thrush, mostly rufous-brown, which is best distinguished by its white lower belly and vent, and black bill. It can be found foraging at all levels in rainforest, from the ground, where it feeds on insects and other invertebrates, to the middle and upper levels, where it eats many palm and lauraceous fruits. It also visits secondary growth and fruiting trees in pastures. Males have a rich, attractive song, not too unlike that of the Clay-coloured Thrush.

Where to see This thrush is a common resident at 750m to 1,200m, occasionally higher, on the Caribbean slope. After breeding, it descends to the lower foothills and even as low as sea level.

Clay-coloured Thrush *Turdus grayi* 23.5cm

This thrush is the national bird of Costa Rica, where it is called *Yigüirro*. It is a very familiar bird. As its name suggests, it is a fairly uniform brownish bird, its dull appearance relieved by a bright yellow bill. It has a beautiful song which is heard mainly before and during the breeding season. It often forages on the ground, and its varied diet includes earthworms, slugs, insects and many kinds of fruit.

Where to see It is resident countrywide from the lowlands up to above 2,000m. It is common in cultivated land of all kinds, including pasture with scattered trees, parkland and suburban gardens.

Elegant Euphonia *Euphonia elegantissima* 11cm

This is an exquisite euphonia; both sexes are distinguished from other euphonias by their light blue hood. Otherwise, the male is mainly blue-black and tawny-orange, the female mainly yellowish olive. Like other euphonias, this species feeds on fruit, specialising on mistletoes from which it removes the skin before swallowing the pulp and seeds.

Where to see During the breeding season it is uncommon to rare between 1,300m and 2,000m on the Pacific slope. Outside the breeding season, most birds move to the Caribbean slope, sometimes down to as low as 750m. The Elegant Euphonia has become rarer because of intense persecution for the cage-bird trade.

Golden-browed Chlorophonia *Chlorophonia callophrys* 13cm

This is a small, brightly coloured bird with a stubby bill that is very distinctive and easily identified within its range. It has an often heard, low-pitched whistle – *wheeeuu*. It is found in highland cloud forest and edges, as well as isolated trees in clearings and pastures. Outside the breeding season, it forages in small flocks at all levels including the canopy, feeding mainly on mistletoe fruits, soft ericaceous berries and a few small figs, as well as small insects and spiders.

Where to see The Golden-browed Chlorophonia is a common resident in the highlands of both slopes, from about 900m up to the timberline on the Caribbean slope and from 1,500m upwards on the Pacific slope. It may descend much lower during the wet season when fruit is scarce at high elevations. It is endemic to Costa Rica and Panama.

Yellow-throated Euphonia *Euphonia hirundinacea* 11cm

The Yellow-throated Euphonia is one of several blue-and-yellow euphonias that occur in Costa Rica. The male is distinguished from most others by lacking a blue-black throat and from Thick-billed by having a dark crown, isolating a yellow forehead patch. The female is olive-green above and yellow below, fading to white on the belly. This euphonia occurs in gallery forest, on the forest edge, and in second growth, plantations and gardens. It eats mostly berries, especially mistletoes, and some insects.

Where to see This euphonia is common on the Pacific approaches to the northern cordilleras and on the Nicoya Peninsula but is rare further south. On the Caribbean slope it is common in the Río Frío area. Its altitudinal range is sea level to about 1,400m.

Thick-billed Euphonia *Euphonia laniirostris* 11cm

Male Thick-billed Euphonias resemble the Yellow-throated Euphonia but have a much bigger yellow patch on the crown and lack white on the belly. Females have a pale yellow lower belly (mainly greyish white in the Yellow-throated Euphonia). This species is often seen in small groups on the forest edge and in cleared areas with scattered trees and gardens, where it forages for epiphytic mistletoe berries as well as a few melastome berries and small figs.

Where to see This euphonia is a locally common resident in south-western areas adjacent to the Panama border, notably the Golfo Dulce lowlands. It is rare further north.

Olive-backed Euphonia *Euphonia gouldi* 9.5cm

The male Olive-backed Euphonia is glossy green above with a conspicuous yellow forehead and a rufous belly. The female is a less distinctive olive-green above and yellowish below but has a rufous forehead and undertail coverts. Olive-backed Euphonias are often seen in pairs or small groups and regularly accompany mixed-species foraging flocks while searching for and eating a wide range of small fruits and berries, including mistletoes, *Anthurium*, *Trema* and *Cecropia* spikes.

♀

Where to see This species is a common resident in the Caribbean lowlands up to 750m in the foothills and 1,000m in the south-east. In the north it reaches the Pacific side of the Cordillera de Guanacaste through low passes.

♂

Sooty-capped Chlorospingus *Chlorospingus pileatus* 13.5cm

This bird's distinctive head pattern is very different to that of the Common Chlorospingus. Its head is black with a very conspicuous stripe above its brick-red eye, which sometimes appears to be kinked. It forages busily among the epiphytes and foliage of mossy highland cloud forest, both primary and secondary, usually in small scattered flocks. It feeds on insects, spiders and berries, including *Miconia*, *Fuchsia*, ericaceous blueberries and blackberries.

Where to see It is a very common resident in cloud forest above 1,600m in the Cordillera de Tilarán and above 2,000m to the timberline in the Cordilleras Central and de Talamanca. It is endemic to Costa Rica and Panama.

Common Chlorospingus *Chlorospingus ophthalmicus* 13.5cm

This chlorospingus is a largely olive-green bird with a grey-brown head and a whitish belly. The conspicuous white patch behind its eye is diagnostic; the similar Ashy-throated Chlorospingus *C. canigularis* (not illustrated) lacks this and has more white on the belly. The Common Chlorospingus is a noisy, active bird that feeds at many levels, but chiefly low, frequently joining mixed flocks of other small birds foraging for insects and many spiders. It has the habit of mashing small fruits in its bill and discarding large seeds and husks.

Where to see This species is very common in wet highland forest above 400m on the Caribbean slope and 1,100m on the Pacific slope. It is rare above 2,300m.

Stripe-headed Sparrow *Peucaea ruficauda* 18cm

This is one of the larger sparrows, with black-and-white stripes on its head. The upperparts are buff with the back and wings streaked with black, while the underparts are white to pale grey. It inhabits relatively open, scrubby areas and the edges of woodland, where it is seen in small flocks that forage on the ground for seeds, supplemented with a few small insects and spiders.

Where to see This sparrow is a common resident in the dry, north Pacific lowlands and foothills, up to 800m on the slopes of the Cordillera de Guanacaste and the hills of the Nicoya Peninsula.

Black-striped Sparrow *Arremonops conirostris* 16.5cm

This chunky sparrow has black stripes on its head, an olive-green back and a yellow patch on its shoulder. The sexes are alike. It is similar to the smaller Olive Sparrow *A. rufivirgatus* (not illustrated), which occurs only in deciduous woodlands on the northern Pacific slope, so there is no overlap in their ranges. The Black-striped Sparrow forages on the ground on the forest edge and in second growth and gardens. It finds insects, spiders and seeds on the ground but also takes insects and picks berries from low vegetation. It is seen in pairs, never in flocks, and is a shy and retiring species.

Where to see The Black-striped Sparrow is a common resident in the humid lowlands and foothills of the Caribbean slope and the south Pacific slope as far north as Carara, from sea level up to about 1,500m. It is absent from the dry north-west.

Orange-billed Sparrow *Arremon aurantiirostris* 15.5cm

The sexes are alike and their bright orange bill and head pattern makes them easily identifiable. This species is usually encountered on the ground in the dark understorey of wet forest where it forages in pairs or family groups for small insects, spiders and berries that have dropped to the ground.

Where to see This species is a common resident in rainforest throughout the Caribbean lowlands and foothills up to about 800m. It also occurs on the south Pacific slope, as far north as Carara National Park.

Chestnut-capped Brushfinch *Arremon brunneinucha* 18.5cm

The head pattern of this species is distinctive with its chestnut cap, black face, white throat and three small white spots on the forehead. There is a black breast-band above the grey belly and the back is dark olive-green. It forages mostly on the ground in the understorey of wet highland forest and shady ravines. It flicks aside leaf litter in search of cockroaches, beetles, spiders, small centipedes and other invertebrates. It also gleans insects and berries from low foliage.

Where to see It is a common but inconspicuous resident of wet rainforests at mid-elevations on both slopes, from about 1,000m up to 2,500m.

Sooty-faced Finch *Arremon crassirostris* 16cm

The Sooty-faced Finch has a distinctive dark grey face, white moustachial stripes and a rufous cap. Otherwise, it is mostly dark olive with a contrasting yellow belly. The sexes are alike. The Sooty-faced Finch is usually found in pairs or small family groups in wet forest in hilly terrain, foraging in ravines or beside streams. It scratches through leaf litter or rummages low down in tangled foliage, taking insects, spiders, and fallen fruits and seeds. Occasionally, it follows army ants.

Where to see This is a locally common resident on the Caribbean slope, mostly from 600m up to 1,500m, from the Cordillera de Tilarán south to Panama. It is endemic to Costa Rica and Panama.

Rufous-collared Sparrow *Zonotrichia capensis* 13.5cm

This is an attractive sparrow with black stripes on its head, a short crest, white throat, black patch on its breast, and a rufous collar that make it easily recognisable. It is commonly found in open and semi-open areas, including villages, farmland, gardens and parks as well as scrubby secondary growth. Males sing from high, open perches – several clear whistles, followed by a trill. Rufous-collared Sparrows are seen in pairs or small flocks, foraging on the ground in search of seeds and occasional small insects.

Where to see The Rufous-collared Sparrow is an abundant resident throughout the foothills and highlands of both slopes, at altitudes of 400m to 3,000m.

Large-footed Finch *Pezopetes capitalis* 20cm

This is a large, dark finch which has a black head with slaty-grey stripes on the crown. It is otherwise mostly olive-green. It is a ground-dwelling bird, often found in pairs or small family groups that scratch in leaf litter. The sexes are alike. Males have a loud song that is often heard at dawn. The Large-footed Finch really does have large feet and uses them to scratch the ground vigorously, both feet at once, in search of seeds and insects.

Where to see This species is found in forest understorey, bamboo thickets and brushy pastures at altitudes above 2,000m the length of the country. It is a regional endemic, found only in the Talamancan montane forests of Costa Rica and Panama, and the Serranía del Darién on the border between Costa Rica and Colombia.

Yellow-thighed Brushfinch *Atlapetes tibialis* 18.5cm

The conspicuous bright yellow, puffy feathers on its thighs make this finch unmistakable. Otherwise, it is dark grey with a mostly black head, wings and relatively long tail. The sexes are alike. It is usually seen in pairs or in small, noisy family groups, which often join mixed-species flocks. The Yellow-thighed Brushfinch is a common bird in wet mountain forests, bamboo clumps, scrubby pasture and bushy clearings. It feeds at all levels from the treetops to the ground, taking insects, spiders and many berries. It also squeezes nectar from flowers.

Where to see It is a common resident in the highlands of the Cordilleras de Tilarán, Central and de Talamanca, generally from 1,500m up to the tree line. It is a regional endemic, occurring only in Costa Rica and western Panama.

Red-breasted Meadowlark *Leistes militaris* 16.5cm

This species is sometimes named the Red-breasted Blackbird, although it is not closely related to the Red-winged Blackbird or its relatives. This attractive blackbird is found in open fields, where it often perches conspicuously on tall grasses or small bushes. The male is an unmistakable black with a bright red breast and throat. The female is brown and streaky but the red tinge on the breast is diagnostic. Listen for the male's song – several metallic chips followed by a raspy buzz. Red-breasted Meadowlarks forage in open, wet areas where they take invertebrates, seeds and occasionally fruit. They are gregarious, particularly in the non-breeding season.

Where to see This is a locally common resident in the Golfo Dulce region. It was first seen in Costa Rica in 1974 and continues to expand its range northwards.

Montezuma Oropendola *Psarocolius montezuma* ♂50cm ♀38cm

The Montezuma Oropendola is the larger of the two Costa Rican oropendolas and it also differs in having a chestnut back, pale blue skin forming a cheek-patch, and an extensive orange tip to the bill. Males are bigger than females. The Montezuma Oropendola breeds in colonies of up to 50 pairs that build pendulous nests close together in tall, isolated trees. The song of the male is a bubbling sound followed by loud gurgles, likened to water being poured from a bottle. It is performed during a mating display when the male bows or swings upside down from a branch. Flocks forage in the forest canopy, where they take small lizards and frogs as well as invertebrates, fruits and nectar.

Where to see This species is a common resident in the Caribbean lowlands, locally as high as 800m.

Chestnut-headed Oropendola *Psarocolius wagleri* ♂35cm ♀28cm

This is a rather large, dark chestnut oropendola with a yellow tail that occurs in rainforest and on the forest edge in the lowlands. It breeds in colonies of up to 12 to 50 pairs, their pendulous nests usually hanging in the crown of a tall, isolated tree. Males are much larger than females but are similar in appearance. These oropendolas forage in noisy flocks, eating many different fruits as well as large insects, spiders, frogs and lizards. They also take nectar from large flowers.

Where to see This oropendola is a common resident in forested areas in the Caribbean lowlands, commonly up to 800m, sometimes as high as 1,700m.

Streak-backed Oriole *Icterus pustulatus* 19cm

The streaked back of this species is diagnostic. Also note the extensive white edging to the wing feathers. The Streak-backed Oriole is found in the canopy and edges of deciduous forest, plantations, gardens and semi-open areas with scattered trees. It forages for caterpillars, beetles and other insects. It also takes more fruit than most other orioles and visits flowering trees to take nectar, as well as to hunt visiting wasps and bees.

Where to see This species is usually the commonest resident oriole in the lowlands and foothills of the north Pacific coast. It occurs south to around Cañas and up to about 450m.

Baltimore Oriole *Icterus galbula* 18cm

♂

The male is the only orange oriole in Costa Rica that has a black head and broad white wing-bars. The female differs in being olive above, tinged with orange, but also has two white wing-bars. The Baltimore Oriole is found in open woodland, cacao and coffee plantations, and scattered trees in farmland and villages. It consumes a lot of nectar, especially of flowering trees, vines and epiphytes such as *Erythrina*, *Inga*, *Calliandra* and *Norantea*. It also forages for insects, spiders and fruits.

Where to see This oriole is a common to abundant migrant and winter resident from sea level to 1,500m, sometimes to 2,000m. It is most numerous in Guanacaste.

Spot-breasted Oriole *Icterus pectoralis* 21cm

This species is easily distinguished from other yellow and black orioles by the distinctive, diagnostic spotting on either side of its breast and white wing-patches on the tertials and base of the primaries. The sexes are alike. The Spot-breasted Oriole lives in open woodland, on the forest edge and in grassland with scattered trees, and is also found around villages and farms. It forages for insects and often visits flowers including *Caesalpinia*, *Erythrina* and *Gliricidia* for nectar, and can often be seen in small flocks, especially outside the breeding season.

Where to see It is an uncommon and local resident in the lowlands and up to about 500m in the foothills of the north Pacific slope. It is most easily seen around the Gulf of Nicoya.

Black-cowled Oriole *Icterus prosthemelas* 19cm

This species is an attractive black-and-yellow oriole with yellow on the shoulders, belly and from the rump up to the lower back. The Black-cowled Oriole forages, gleans and probes for food actively in forest borders, plantations, along streams and also in more open habitats, taking insects, nectar from flowering trees, and berries. The sexes are alike in Costa Rica, though not in Mexico and northern Central America.

Where to see This oriole is a fairly common resident throughout the Caribbean lowlands, up to 700m locally and occasionally higher to 1,200m.

Red-winged Blackbird *Agelaius phoeniceus* 22cm

Black males, with their red and yellow shoulder patches, are easy to identify. Females have a longish, sharply pointed bill and are heavily streaked below, which distinguishes them from other species in the same habitats. Large flocks of Red-winged Blackbirds are seen in the non-breeding season, foraging in pastures and agricultural fields, seeking insects and seeds. They are sometimes a pest of newly sown grain fields.

Where to see This species is a locally common to abundant resident in Guanacaste in the Tempisque and Bebedero Basins. It also occurs in the Río Frío region of the Caribbean slope.

Melodious Blackbird *Dives dives* 25cm

This is a distinctive species even though it lacks obvious field marks. Its plumage is wholly black with a blue sheen, both above and below. It also has a sharply pointed bill and dark eyes. The Melodious Blackbird uses a wide range of habitats, but avoids dense forest and thick undergrowth. It has adapted to human habitation and is readily seen in gardens. The Melodious Blackbird forages mainly on the ground for insects but will also take nectar and ripening maize ears as well as many different fruits.

Where to see This species has benefited from deforestation, which has allowed it to rapidly expand its range south along the Pacific slope of Central America through Nicaragua and onwards. Prior to 1989 there was only one Costa Rican record. It is now commonly seen in semi-open habitats in the tropical lowlands along much of the Pacific slope.

Great-tailed Grackle *Quiscalus mexicanus* ♂43cm ♀33cm

The male has yellow eyes but is otherwise glossy purplish black with a long tail that is often held cocked. The female is basically brown with a pale stripe above the eye and a shorter tail. The Great-tailed Grackle is a gregarious species that forages mostly on the ground in agricultural areas and is more or less omnivorous, eating carrion, small vertebrates, birds' eggs and nestlings, large insects, grain and miscellaneous fruits.

♀

Where to see This grackle's range has been spreading since the 1960s and it is now found countrywide in non-forest habitats, particularly in agricultural areas and villages up to about 1,500m.

♂

Northern Waterthrush *Parkesia noveboracensis* 13.5cm

The Northern Waterthrush is a relatively large, terrestrial warbler with a whitish supercilium and heavy streaking on its white underparts, which often have a yellow wash. It walks rather than hops and mostly stays on the ground, constantly teetering and bobbing its rear end. The Northern Waterthrush is found in a variety of waterside habitats – shady mountain rivers and streams with rocky margins, wooded swamps and mangroves, where it forages for small aquatic invertebrates.

Where to see This waterthrush is a common passage migrant and winter resident from the lowlands up to 1,500m.

Golden-winged Warbler *Vermivora chrysoptera* 11.5cm

This beautiful warbler is distinctive if seen well and the bright yellow bars on its wings are diagnostic. Unfortunately, it is declining over much of its range. It also hybridises with the Blue-winged Warbler *V. cyanoptera* (not illustrated). The female is duller than the male. It is usually found in forested areas, especially in the forest canopy, tall second growth and sometimes forest clearings or gardens. It often accompanies other small birds in mixed-species flocks, feeding on small insects and other invertebrate prey.

Where to see This is a widespread migrant and winter visitor in the lowlands and middle elevations on both slopes. It is most numerous at 700m to 1,400m on the Caribbean slope.

Tennessee Warbler *Leiothlypis peregrina* 11.5cm

This is a rather plain warbler with olive-green upperparts and white to pale yellow underparts. It has a pale supercilium and dark eye-stripe, and some have a pale wing-bar. Males have a grey head after moulting into breeding plumage. They are usually seen in semi-open habitats, plantations and gardens, where they forage for both insects and nectar.

Where to see This species is a common passage migrant during both fall and spring migration. It is also a common to abundant winter resident from sea level up to 2,300m on both slopes, but is most numerous at middle elevations.

Grey-crowned Yellowthroat *Geothlypis poliocephala* 13.5cm

The male's black mask is much smaller than on other species of yellowthroat, and the only black on the female's face is restricted to a small patch in front of the eye. The male has a grey cap and is olive-green above and yellow below. The female is generally browner except for her yellow throat. These yellowthroats forage for insects and other invertebrates in meadows, savannahs, sugarcane and other crops, and also take some berries.

Where to see Grey-crowned Yellowthroats live in the lowlands of both slopes up to about 1,500m or higher. Due to deforestation in recent years, this is a species that has increased greatly on the Caribbean slope and in the Golfo Dulce region.

Mangrove Warbler *Setophaga petechia* 11.5cm

The Mangrove Warbler is similar to the American Yellow Warbler *S. aestiva* (not illustrated), but is a little larger and the male's head and throat are rufous-chestnut, not yellow. It was formerly considered to be conspecific with the American Yellow Warbler but is ecologically and behaviourally very distinct. As its name suggests, it is confined to mangroves and their close vicinity, foraging like other warblers for small insects.

Where to see This is a common resident of mangroves along Costa Rica's Pacific coast, from the Gulf of Nicoya to the Panama border.

♂

♀

Blackburnian Warbler *Setophaga fusca* 11.5cm

In all plumages, the Blackburnian Warbler has a triangular darkish cheek-patch framed by yellow or orange, and two white wing-bars. Females and immatures are duller versions of the male. It is usually seen in the forest canopy, on the forest edge, or in secondary forest but also visits gardens. It forages for small insects and spiders and often hovers or pursues prey in flight. It also joins mixed-species foraging flocks.

Where to see This is a common passage migrant and winter visitor, mainly at middle elevations between 900m and 1,500m but locally from the lowlands up to 2,500m.

♀

♂

Chestnut-sided Warbler *Setophaga pensylvanica* 11.5cm

The Chestnut-sided Warbler has two yellowish wing-bars in all plumages and usually a visible trace (more obvious in males) of a chestnut band on its flanks. Chestnut-sided Warblers are common winter visitors in Costa Rica, foraging in forest, but also venturing into clearings and gardens, where they glean small insects, caterpillars and spiders. They also take some berries.

Where to see The Chestnut-sided Warbler is a common passage migrant countrywide, from the lowlands up to 1,850m, and a common winter resident on both slopes (except for the dry north-west) up to 1,500m.

♀

♂

Black-throated Green Warbler *Setophaga virens* 12cm

This is a lively, attractive warbler with a bright yellow face, dark streaks on the white flanks and a yellowish vent. It is found most often in the forest canopy and edges but also visits second growth and gardens. It is an active forager and frequently hovers to pick insects off the undersides of leaves and search the upper surfaces of foliage. It hawks flying insects and chases between trees after flying prey more often than other warblers. It also takes berries and, on its wintering grounds, will often eat the protein corpuscles of tropical *Cecropia* trees.

♂

Where to see This warbler is common in the Caribbean lowlands while on migration and is also a winter visitor, mostly between 1,000m and 3,000m in the Cordilleras Central and de Talamanca, but lower in the northern cordilleras.

Buff-rumped Warbler *Myiothlypis fulvicauda* 13cm

As its name suggests, the rump and much of the tail of the Buff-rumped Warbler is pale buff, made very conspicuous by the contrast with the rest of its plumage and the constant fanning and flirting of its tail. It forages by the side of shady, forested rivers and streams, catching small invertebrates at the water's edge, or often sallying to catch insects in flight. It has a very loud song in the form of a series of accelerating *chip* notes.

Where to see The Buff-rumped Warbler is a common resident throughout the lowlands of the Caribbean and south Pacific slopes. It occurs up to about 1,100m on the Caribbean slope, and 1,500m on the south Pacific slope.

Golden-crowned Warbler *Basileuterus culicivorus* 12cm

The Golden-crowned Warbler is a small, lively species that is common in rainforest in the lowlands and foothills where it often accompanies mixed-species feeding flocks. As with many other tropical warblers, the sexes are alike – greyish above and yellowish below with blackish head stripes that border an easily overlooked golden central crown-stripe. It forages for small insects and spiders, as well as various small fruits.

Where to see This warbler is resident in mountain forests throughout, between 300m and 1,500m on the Caribbean slope and 900m and 2,150m on the Pacific slope.

Black-cheeked Warbler *Basileruterus melanogenys* 13.5cm

This is a small, greyish bird with a black face, bold white supercilium and rufous cap. A clear view of the head should eliminate any confusion with other understorey warblers. The sexes are alike. The Black-cheeked Warbler is normally found in oak forests with a dense bamboo understorey. It often forages for small insects with mixed-species flocks.

Where to see This warbler is a resident breeding bird, endemic to the Talamancan montane forests of Costa Rica and western Panama. It can be seen in the Cordilleras Central and de Talamanca from 1,600m up to the timberline and sometimes in páramo above.

Black-eared Warbler *Basileuterus melanotis* 13cm

This warbler has an unusual pattern of blackish stripes on its head. It also has a usually concealed orange central crown-stripe. Otherwise it is not very distinctive, with dull olive-green upperparts and pale yellow underparts. It tends to forage in small family groups, members calling back and forth to each other, and adults often associate with mixed flocks. This warbler is most easily found in the forest understorey, in both primary and mature secondary forest, including at their edges.

Where to see It is common in the wet middle elevations, between 1,000m and 2,200m, mostly on the Caribbean slope from the Cordillera de Tilarán south to Panama. it is endemic to Costa Rica and Panama.

Wilson's Warbler *Wilsonia pusilla* 11cm

The black cap of the male Wilson's Warbler distinguishes it from other small yellow warblers. The female is similar but lacks the black cap, so is more difficult to distinguish from some other female warblers. This warbler is a very active species that flits about in the canopy, clearings and on the forest edge in search of small insects. It often joins mixed-species foraging flocks.

Where to see This species is a common to abundant migrant and winter visitor, except in the Pacific lowlands where it is rare and irregular. It mainly winters above 900m; it becomes commoner up to the timberline and even occurs in páramo.

♂

Slate-throated Whitestart *Myioborus miniatus* 12cm

This distinctive warbler has long bristles at the base of its broad flat bill. The rufous-red crown that is visible in the lower photograph is not always exposed. On the other hand, its outer tail feathers are tipped with white and are often displayed when the tail is fanned. This whitestart flits around actively in the middle levels of highland forest, often behaving like a flycatcher with acrobatic aerial chases after flying insects. It also forages in clearings, gardens and along hedgerows in agricultural areas. It regularly feeds on the protein corpuscles of *Cecropia* trees.

Where to see The Slate-throated Whitestart is a common resident throughout mountainous areas above 750m on the Caribbean slope and above 1,100m on the Pacific slope, ranging up to 2,150m, except in drier areas.

Collared Whitestart *Myioborus torquatus* 12.5cm

The Collared Whitestart's bright yellow face and rufous crown make it a distinctive bird. Its back, wings and tail are grey to black, and its belly is yellow with a dark grey breast-band. The tail feathers have white edges that are displayed whenever the tail is fanned (which happens often). The Collared Whitestart is easily found in cloud forest and adjacent areas, including hedges and pastures, and often forages with mixed-species flocks in search of its insect prey.

Where to see It is a common resident in the mountains throughout, occuring above 1,600m in the Cordillera de Tilarán, and from 1,850m up to the timberline in the Cordilleras Central and de Talamanca. It is endemic to Costa Rica and Panama.

Summer Tanager *Piranga rubra* 16.5cm

The male has a prominent thickish, pale bill and otherwise is entirely red, a little darker on the wings. The female is olive-brown above and yellow-olive below with a hint of ochre, brightest on the undertail coverts. The female also has a hefty, pale bill, though the upper mandible is darker than the male's. Summer Tanagers are winter visitors that occur in various habitats, including forest canopy, second growth and gardens, feeding on fruit and insects. They tend to be bee and wasp specialists, catching them in flight and beating and rubbing them against a branch to kill them and remove the sting.

Where to see The Summer Tanager is a common migrant and winter resident from about mid-September to late April. It occurs on both slopes from the lowlands up to 2,500m or more.

♀

♂

Tooth-billed Tanager *Piranga lutea* 18cm

The Tooth-billed Tanager is large and stocky. Both sexes are very similar to the respective sex of Summer Tanager, but differ in the heavier, dark grey bill with a protrusion along the cutting edge (hence 'Tooth-billed'). The male Tooth-billed is also not as bright red as the Summer. The Tooth-billed Tanager is found in the canopy and edge of forest in mountainous areas where it forages for insects and fruits. The latter include many different berries and arillate seeds. It also joins mixed-species flocks of other tanagers, warblers and other small birds.

Where to see The Tooth-billed Tanager is a common resident at mid-elevations the length of the country on both slopes.

♂

♀

Black-cheeked Ant Tanager *Habia atrimaxillaris* 18cm

This tanager is a generally dark grey bird with a blackish face and a wash of pink on its underparts, brightest on the throat. The sexes are more or less alike. The Black-cheeked Ant Tanager often joins mixed-species flocks as it moves through the forest, searching for insects by rummaging through dead leaves and picking insects and berries from vegetation. It also accompanies army ants.

Where to see This species is endemic to the Golfo Dulce lowlands of Costa Rica. It is becoming scarcer as its habitat is destroyed, though it is still fairly common in Corcovado National Park and in other rainforest that remains on the Osa Peninsula and around the Golfo Dulce.

Black-thighed Grosbeak *Pheucticus tibialis* 20cm

This grosbeak is a mostly yellow bird with a massive bill, a small black mask and black wings, back, thighs and tail. It has a small but conspicuous white wing patch. The sexes are basically similar. It favours exposed perches in forest and semi-open habitats. The Black-thighed Grosbeak is found in the canopy and edge of montane forest and pastures with scattered big trees. It forages for fruits, seeds and insects, often catching the latter in flight.

Where to see This species is resident in the highlands of both slopes, from 1,000m up to 2,600m on the Pacific slope and 1,500m to 2,600m on the Caribbean slope. It is endemic to Costa Rica and Panama.

Black-faced Grosbeak *Caryothraustes poliogaster* 16.5cm

This bird has a heavy bill, slightly hooked at the tip, and (as its name suggests) a black face. Otherwise it has an olive-green back and wings and a mustard-yellow head and breast. The sexes are alike. The male's calls are buzzy splutters and whistles. This species is often seen in noisy groups, occasionally of 20 or more birds, moving through the middle and upper levels of humid forest or on the forest edge, foraging for insect, nectar, fruits and arillate seeds.

Where to see The Black-faced Grosbeak is a common resident in the lowlands and foothills of the Caribbean slope up to about 900m. It is absent or rare in the drier areas south of Lake Nicaragua.

Blue-black Grosbeak *Cyanoloxia cyanoides* 16.5cm

Males often look simply dark but in good light they are bright violet-blue. Females are very different – mostly rich, dark brown, slightly paler below. Both sexes have a massive, dark, conical bill. Blue-black Grosbeaks are found in the understorey of humid rainforest, dark ravines and shady semi-open meadows and grain fields. Their main diet is seeds (which are crushed before being consumed) and fruit. They also glean for insects, usually in pairs but occasionally as part of a mixed-species flock.

Where to see The Blue-black Grosbeak is a common resident of the Caribbean and south Pacific slopes, from sea level up to 1,200m, mostly in humid evergreen forest, gallery forest or wet ravines.

Red-legged Honeycreeper *Cyanerpes cyaneus* 11.5cm

In breeding plumage, the male is mainly blue with black wings and red legs; in flight, the underside of the wings are bright sulphur-yellow. After the breeding season, the male moults into an eclipse plumage, which is mainly greenish with black wings. Females and immatures are mainly green, with faint streaking on the underparts. The female has red-brown legs, the immature brown. The Red-legged Honeycreeper is found in forest and on the forest edge, venturing lower in scrub and gardens. It often occurs in flowering trees, using its long bill to probe flowers for nectar. It also consumes small fruits and insects.

Where to see This species is a common resident in the dry northern lowlands and is fairly common from sea level to 1,200m on the Pacific slope south to Panama.

Green Honeycreeper *Chlorophanes spiza* 13cm

The predominant colour of the male is a bright, glossy blue-green which is darker on the wings and tail. Males also have a black cap and face, and a red iris. The bill is slightly downcurved with a dark upper mandible and yellow lower mandible. The female is less striking – wholly dullish grass-green with paler underparts. Green Honeycreepers forage in the forest canopy but also descend to the forest edge, clearings with trees and shrubs, and gardens. They often join mixed-species flocks and feed on small fruits, arillate seeds, nectar and small insects.

Where to see This species is resident on the Caribbean slope from sea level up to 1,000m but is most often seen above 300m. It is also common in the lowlands and foothills of the south Pacific slope up to 1,200m.

♂

♀

Shining Honeycreeper *Cyanerpes lucidus* 10cm

All plumages of this species have a more strongly downcurved bill and a shorter tail than the more common Red-legged Honeycreeper. Also, note the characteristic yellow legs. The male is deep blue overall with a black throat and wings. The female is less distinctive but has a necklace of bluish streaks across the breast.

The Shining Honeycreeper is found in humid evergreen forest and edge. It feeds at all levels, but mostly in the canopy of fruiting and flowering trees and bushes. It eats many berries and arillate seeds, nectar from diverse flowers, and small insects and spiders that it gleans from vegetation.

Where to see This honeycreeper is a locally common to abundant resident of the lowlands and foothills on both the Caribbean slope and south Pacific slope, from sea level to 1,200m.

Scarlet-thighed Dacnis *Dacnis venusta* 12cm

The male Scarlet-thighed Dacnis is turquoise-blue above and black below with sometimes inconspicuous scarlet thighs, red eyes and a very sharp, pointed bill. The female is less distinctive but has a turquoise face and a buffy belly. It is usually found in forest canopy and on the forest edge, and in secondary forest, semi-open areas and gardens. It often joins mixed flocks, foraging for diverse small fruits, insects and spiders.

Where to see The Scarlet-thighed Dacnis is an abundant resident on the Caribbean slope and south Pacific slope as far north as Carara. Its altitudinal range extends from the lowlands up to 1,500m, sometimes a little higher.

Buff-throated Saltator *Saltator maximus* 20cm

This species can be distinguished by its rather heavy bill, its buff throat broadly bordered by black, its conspicuous white supercilium and its olive-green back. It stays paired throughout the year and is found alone or in pairs on the forest edge, or in overgrown pastures, brushy plantations and gardens. It often joins mixed-species flocks and forages for insects, spiders and other invertebrates, many fruits, and nectar that it squeezes from flowers with its bill.

Where to see This saltator is resident, rare and local in the dry north-west, but otherwise common to abundant countrywide from sea level up to 1,200m, sometimes higher.

Bananaquit *Coereba flaveola* 9cm

This is a small, active bird with a characteristically short, sharply pointed, downcurved bill. Its plumage is mostly dark grey above, yellow below, with a conspicuous white stripe above the eye and a grey throat. It feeds on nectar, often by piercing the base of large flowers, and also forages acrobatically in foliage, searching for insects and juicy fruit. It can be seen in a wide range of wooded habitats and often joins mixed-species flocks of foraging birds.

Where to see The Bananaquit is a common resident throughout the Caribbean slope, and from Carara south on the Pacific slope, from sea level up to about 1,500m.

Blue-black Grassquit *Volatinia jacarina* 10cm

The male is blue-black all over in its breeding plumage. The female is largely brown, darker above, paler below and streaked. Moulting and immature males are patchy blue-black and brown. They live in open, grassy areas, weedy fields and farmland where they tend to forage inconspicuously in the low vegetation. The male is more conspicuous when he makes repeated song-jumps in which he simultaneously flits up from a perch and utters short, buzzy phrases, before fluttering back to the same perch. Blue-black Grassquits feed chiefly on seeds but also take some small fruits and invertebrates.

Where to see The Blue-black Grassquit is common throughout the country in the lowlands and foothill from sea level up to about 1,700m.

Grey-headed Tanager *Eucometis penicillata* 16cm

The grey head of this tanager, with its short, bushy crest, is diagnostic. The rest of its upperparts are olive-green and its underparts are a rich yellow, sometimes tinged ochraceous. The sexes are alike. The song is a rich, squeaky warble, at times prolonged. The Grey-headed Tanager typically inhabits the low to mid-levels of shady forest understorey, occurring as singles, pairs or family groups. It often follows army-ant swarms, catching fleeing insects and spiders. It also gleans insects from foliage and takes many different small fruits.

Where to see It is an uncommon resident in the dry north-west lowlands, where it is confined to moister areas, but it is very common on the Nicoya Peninsula and lower slopes of the Cordillera de Guanacaste. It is also common throughout the lowlands and foothills of the south Pacific slope, up to 1,200m.

White-lined Tanager *Tachyphonus rufus* 17cm

The males and females of this tanager are strikingly different. The male is all black except for a white shoulder patch that is just a sliver and rarely visible. The female is wholly cinnamon-brown with no additional distinctive features. This species is commonly seen in pairs and rarely joins mixed-species flocks. It is not a forest bird, but prefers more open areas of woodland, young second growth and shrubby edges. It feeds on insects and many fruits, typically squeezing out and swallowing fruit pulp and small seeds, then discarding the husks.

♀

Where to see This is a fairly common resident of the Caribbean slope. It has spread to the Osa Peninsula lowlands in recent years and continues its range expansion on both slopes as deforestation provides new opportunities.

♂

Scarlet-rumped Tanager *Ramphocelus passerinii* 16cm

The male is unmistakable with its brilliant red lower back and rump, contrasting with the velvety black of the rest of the plumage and the distinctive silver-blue bill. The female has a grey-brown head, an olive-brown back and an olive-yellow breast and rump. On the Pacific slope, the subspecies *R. p. costaricensis* (previously considered a separate species, Cherrie's Tanager) may be seen, identifiable by the female's orange breast and rump. The Scarlet-rumped Tanager frequents semi-open areas, including gardens, where it feeds on many fruits and berries, including *Cecropia* and *Piper* spikes, and diverse insects and spiders.

Where to see This tanager inhabits the lowlands up to mid elevations on both slopes.

Variable Seedeater *Sporophila corvina* 10.5cm

There are two subspecies of Variable Seedeater in Costa Rica. The male on the bottom left is the Caribbean subspecies *S. c. corvina*, which is almost all black except for white wing-linings, small white wing-bars on its primaries and a narrow white line in the centre of its belly. The Pacific subspecies male (*S. c. hoffmanni*, top) has a white collar, and considerably more white on its underparts and rump. In both forms, females are olive-brown with white wing linings. The Pacific subspecies is shown (bottom right); it has a paler belly, while the Caribbean is more uniform olive-brown below. Variable Seedeaters are birds of semi-open grasslands, gardens and forest edges, feeding on seeds, berries and a few insects, often in mixed-species flocks.

Where to see This species is common throughout the lowlands and foothills of the Caribbean slope and the southern Pacific slope, up to about 1,500m and north to around Carara.

♂

♂

♀

Peg-billed Finch *Acanthidops bairdii* 13.5cm

The male Peg-billed Finch is dull grey, slightly paler below. The female is olive-brown, paler below and faintly streaked, with brown wing-bars. Both sexes have a distinctive, long, very slightly upturned, bicoloured bill. The Peg-billed Finch is found in forest edge with thickets of bamboo, where it forages, sometimes in mixed flocks, for berries, bamboo and grass seeds, nectar and small insects. It is almost always found around seeding bamboo.

Where to see This is an uncommon to sometimes common (possibly nomadic) resident in the highlands, up to 1,500m in the Cordillera de Tilarán and up to the timberline on the Cordilleras Central and de Talamanca.

♂

Slaty Flowerpiercer *Diglossa plumbea* 10cm

The male Slaty Flowerpiercer is dull bluish grey, the female browner. The unusual up-tilted bill gives this species a distinctive appearance. The hooked upper mandible is used to grip flowers while the lower mandible pierces the corolla, making a hole through which the flowerpiercer's brush-tipped tongue can extract nectar. It also gleans small insects from foliage and sometimes catches them in flight.

♂

Where to see This bird is endemic to the Talamancan montane forests of Costa Rica and Panama. It is a common resident in montane forest canopy, forest edges and summit scrub, all areas where flowers are usually abundant. Found up to high elevations throughout – up to 2,100m in the Cordilleras Central and de Talamanca, lower on lower cordilleras.

Speckled Tanager *Ixothraupis guttata* 13cm

The aptly named Speckled Tanager is green above and white below, and densely speckled with black both above and below. The sexes are alike. It is found in pairs or small flocks, often with other tanagers or honeycreepers. It often moves through the canopy and upper levels of humid primary forest, secondary growth and clearings with big trees, searching for fruits, particularly small figs and melastome berries. Speckled Tanagers also take insects, spiders and other invertebrates.

Where to see This is an uncommon to fairly common resident tanager found in foothill forest at 400m to 1,000m on the Caribbean slope from the Cordillera de Tilarán south, and at 300m to 1,400m on the south Pacific slope.

Blue-grey Tanager *Thraupis episcopus* 15cm

This is one of the most familiar and easily identified birds of Costa Rica. The sexes are similar though the male's blue-grey plumage and blue wings are a little brighter. The Blue-grey Tanager eats a great variety of mostly small fruits as well as insects and spiders that are gleaned from foliage. It also takes nectar from large flowers.

Where to see This tanager is a common resident throughout most of the country up to about 2,000m, but is generally rare in the dry north-west, except in humid gardens.

Palm Tanager *Thraupis palmarum* 16cm

The greyish olive-green of this species can have a glossy-grey sheen in some lights, but the black flight feathers are diagnostic. It is usually seen in pairs in fairly open habitats with scattered trees, including parks, gardens and plantations. It feeds on fruits and also small invertebrates, which it gleans from foliage or occasionally catches in flight.

Where to see It can be found throughout the country up to an elevation of 1,500m but is rare in the drier north-west. It is particularly common among coconut palms along the Caribbean coast.

Golden-hooded Tanager *Stilpnia larvata* 13cm

This attractive tanager has a largely yellow-buff head with a black mask edged with blue. The sexes are alike except that females are duller. Golden-hooded Tanagers are birds of the forest canopy, open wooded areas, secondary growth and gardens. They are usually in pairs or small flocks. They eat much fruit, especially figs, melastomes, *Cecropia* and *Piper* spikes, and arillate seeds, but also search mossy branches and foliage for insects.

Where to see This is a common resident in the lowlands and foothills of the Caribbean slope and south Pacific slope (north to Carara). It occurs from sea level up to about 1,500m.

Silver-throated Tanager *Tangara icterocephala* 13cm

This tanager has a distinctive silvery throat but otherwise is mostly lemon-yellow, with black wings and tail and black stripes on its back. The sexes are similar though the female is a little duller. The call is a characteristic buzzy *zzeep*. Silver-throated Tanagers are common and active, usually found in pairs or small flocks, and sometimes accompany mixed-species feeding flocks with other small birds. The Silver-throated Tanager is found in wet mountain forest and cloud forest, as well as semi-open clearings and gardens. It feeds on small insects and a variety of small fruits, including melastomes, elderberries and small figs.

Where to see This tanager is a common to abundant resident at altitudes of 600m to 1,700m in wet mountain forests, plus adjacent semi-open clearings and gardens on both slopes. Sometimes it descends to sea level during heavy rains in the wet season.

Spangle-cheeked Tanager *Tangara dowii* 13cm

This tanager is a beautiful, multi-coloured bird, unique among tanagers with its black face, orange belly, cobalt wings, and the turquoise spangles on its chest, cheeks and nape. The sexes are similar, and it does not resemble any other tanager. The Spangle-cheeked Tanager is common in the canopy of epiphyte-laden mountain forests and in semi-open clearings with trees, second growth and forest edge. It often forages in pairs, family groups or with mixed-species feeding flocks. It feeds on small insects, spiders and diverse fruits.

Where to see This is a common resident in the mountains from the Cordillera de Tilarán south to Panama. It is found at altitudes between 1,200m and 2,750m and, in smaller numbers, as low as 800m or as high as 3,000m.

GLOSSARY

Cere A fleshy patch at the top of the beak, surrounding the nostrils.

Eye-stripe A line of colour (usually dark) running in front of and behind the eye.

Gorget A patch of colour on the throat.

Lek A communal area in which birds gather for display and courtship behaviour.

Lores The area between the base of the bill and the eye.

Malar stripe A stripe (usually dark) running from the base of the beak, separating the cheek from the throat.

Moustachial stripe A stripe (usually dark) running along the lower side of the cheek (above the malar stripe).

Primaries The outer flight feathers, or 'hand' of the wing.

Secondaries The inner flight feathers, or 'arm' of the wing.

Supercilium A line of colour (usually pale) above the eye, in an 'eyebrow' position.

Tertials Wing feathers closest to the body.

Traplining/trapliner A feeding strategy whereby an individual visits known food sources in a repeated order on a regular basis. / A bird that engages in traplining.

Undertail-coverts A set of feathers on the underside of the body that cover the base of the tail feathers.

Vent
The external opening of the cloaca, hidden by feathers.

PHOTO CREDITS

Key: t = top; tl = top left; ti= top inset; c = centre; b = bottom; bl = bottom left; br = bottom right; bi = bottom inset.
Photo agency abbreviations: SS = Shutterstock; iS = iStock.

1 Martin Mecnarowski/SS; **3** artiste9999/iS; **5** Gianfranco Vivi/SS; **8** c CL-Medien/SS; b Ondrej Prosicky/SS; **9** Ronald Smijers/SS; **10** b Agami Photo Agency/SS; **20** t Michel VIARD/iS; **23** t Danita Delimont/SS; **24** t AGAMI; **29** tl Ondrej Prosicky/SS; b Murray Cooper; **32** b AGAMI; **40** bl Cavan Images/iS; br AGAMI; **41** t FotoRequest/SS; br crbellette/SS; **42** bl Jim Cumming/SS; **46** tr NNehring/iS; **48** b Jaime Espinosa/SS; **49** b David Havel/SS; **50** t Joe Beck; **51** t David Havel/SS; **55** t FalcoWildlifePhoto/Flickr; **59** t Jalil El Harrar/Nature Picture Library; **61** b AGAMI; **65** b Allen Lara Gonzalez/Shutterstock; **77** b Rocky Cranenburgh/SS; **79** t Pascale Gueret/SS; **87** t Traveller MG/SS; **85** b James Lowen; **88** t Jim Cumming/SS; **90** t AGAMI; **93** b AGAMI; **104** t Patrick Gijsbers/iS; **110** t Harry Collins Photography/SS; **112** bi Artush/iS; **113** ti Henk Bogaard/iS; **114** bi phototrip/iS; **115** bi Ondrej Prosicky/iS; **116** t Jim Cumming/SS; ti QueGar3/iS; **121** b artiste9999/SS; **127** b Murray Cooper; **132** b Jan Flindt Anderson; **158** t Banu R/iS; b AGAMI; **172** t neil bowman/iS; **190** t Nick Athanas/Tropical Birding Tours; **194** t Pierre Williot/SS; **193** b phototrip/iS; **214** t James Lowen; b Agami Photo Agency/SS; **217** b carlos Pereira/iS.

INDEX

Amazon, Mealy 120
Red-lored 119
White-fronted 119
Anhinga 73
Ani, Groove-billed 56
Smooth-billed 56
Antbird, Bicolored 139
Chestnut-backed 140
Dusky 138
Spotted 139
Zeledon's 141
Antpitta, Ochre-breasted 142
Scaled 141
Spectacled 142
Streak-chested 142
Antshrike, Barred 134
Black-crowned 136
Black-hooded 135
Fasciated 137
Great 137
Antvireo, Plain 133
Antwren, Dot-winged 132
Slaty 133
Aracari, Collared 110
Fiery-billed 109
Attila, Bright-rumped 156

Bananaquit 212
Barbet, Prong-billed 108
Red-headed 108
Barbtail, Spotted 130
Barbthroat, Band-tailed 24
Becard, Barred 164
Black-and-white 166
Cinnamon 165
Rose-throated 166
White-winged 165
Bellbird, Three-wattled 158
Blackbird, Melodious 196
Red-breasted 192
Red-winged 196
Bobwhite, Spot-bellied 17
Brilliant, Green-crowned 33
Brushfinch, Chestnut-capped 189
Yellow-thighed 191

Caracara, Crested 117
Chachalaca, Grey-headed 14
Plain 14
Chlorophonia, Golden-browed 184
Chlorospingus, Ashy-throated 187
Common 187
Sooty-capped 187
Coquette, Adorable 32
Black-crested 32
White-crested 32
Cotinga, Snowy 159
Cuckoo, Lesser Ground 57
Squirrel 57
Curassow, Great 16

Dacnis, Scarlet-thighed 211
Dove, Grey-chested 62
Grey-headed 62
Inca 60
Mourning 63
White-tipped 62
Duck, Muscovy 13

Egret, Great 78
Snowy 80
Western Cattle 77
Elaenia, Mountain 143
Yellow-bellied 143
Emerald, Blue-tailed 43
Canivet's 43
Coppery-headed 46
Garden 43
Euphonia, Elegant 183
Olive-backed 186
Thick-billed 185
Yellow-throated 185

Fairy, Purple-crowned 29
Falcon, Barred Forest 118
Laughing 117
Slaty-backed Forest 88
Finch, Large-footed 191
Peg-billed 217
Sooty-faced 190
Flatbill, Eye-ringed 146
Yellow-olive 147
Flowerpiercer, Slaty 217
Flycatcher, Black-capped 149
Boat-billed 153
Brown-crested 155
Dusky-capped 155
Golden-bellied 151
Grey-capped 150
Northern Scrub 145
Northern Tufted 148
Ochre-bellied 145
Olive-streaked 144
Olive-striped 144
Scissor-tailed 154
Social 149
Streaked 152
Sulphur-bellied 152
Vermillion-crowned 149
White-ringed 151
Yellowish 148
Foliage-gleaner, Fawn-throated 130
Lineated 129
Scaly-throated 128
Frigatebird, Magnificent 72

Gallinule, Purple 64
Gnatwren, Tawny-faced 177
Trilling 177
Goldentail, Blue-throated 55
Grackle, Great-tailed 197
Grassquit, Blue-black 213
Grebe, Least 65
Pied-billed 66
Grosbeak, Black-faced 208
Black-thighed 208
Blue-black 209

Ground-dove, Ruddy 60
Guan, Black 15
Crested 15
Gull, Laughing 71

Hawk, Barred 87
Broad-winged 87
Common Black 86
Grey 89
Mangrove Black 86
Roadside 87
Semiplumbeous 88
Short-tailed 90
Swainson's 90
Tiny 85
White 88
Zone-tailed 90
Hermit, Bronzy 23
Green 26
Long-tailed 25
Stripe-throated 25
Heron, Bare-throated Tiger 75
Boat-billed 76
Green 76
Little Blue 79
Rufescent Tiger 74
Striated 76
Tricolored 78
Honeycreeper, Green 210
Red-legged 209
Shining 211
Hummingbird, Beryl-crowned 54
Black-bellied 47
Blue-chested 53
Blue-vented 49
Charming 54
Cinnamon 51
Fiery-throated 31
Lovely 53
Mangrove 53
Rivoli's 34
Ruby-throated 40
Rufous-tailed 52
Scaly-breasted 48
Scintillant 42
Snowy-bellied 50
Steely-vented 49
Stripe-tailed 47
Talamanca 34
Violet-headed 43
Volcano 41

Ibis, American White 73

Jabiru 72
Jacamar, Rufous-tailed 106
Jacana, Northern 67
Jacobin, White-necked 21
Jay, Azure-hooded 168
Brown 168

Kingbird, Tropical 153
Western 153
Kingfisher, Amazon 101
American Pygmy 101
Green 102
Ringed 103
Kiskadee, Great 150
Kite, Hook-billed 84
Plumbeous 85
Snail 86
Swallow-tailed 84

Lancebill, Green-fronted 27
Leaftosser, Grey-throated 122
Tawny-throated 122
Limpkin 65

Macaw, Scarlet 121
Magpie-jay, White-throated 169
Manakin, Long-tailed 160
Orange-collared 162
Red-capped 163
Velvety 160
White-collared 161
White-rumped 159
Mango, Green-breasted 31
Meadowlark, Red-breasted 192
Motmot, Broad-billed 106
Lesson's 104
Rufous 104
Turquoise-browed 105
Mountaingem, Grey-tailed 38
Purple-throated 37
Variable 38
White-bellied 36
White-throated 38
Mourner, Rufous 154
Myiobius, Sulphur-rumped 163

Nighthawk, Lesser 17
Nightingale-thrush, Black-billed 180
Black-headed 179
Slaty-backed 179
Nightjar, Dusky 18

Oriole, Baltimore 194
Black-cowled 195
Spot-breasted 195
Streak-backed 194
Oropendola, Chestnut-headed 193
Montezuma 192
Osprey 83
Owl, Andean Pygmy 91
Black-and-white 94
Central American Pygmy 91
Crested 93
Ferruginous Pygmy 91
Middle American Screech 92
Mottled 94
Pacific Screech 92
Spectacled 93
Tropical Screech 91

Parakeet, Orange-chinned 118
Orange-fronted 120
Pauraque 18
Pelican, Brown 80
Phainoptila, Black-and-yellow 170
Pigeon, Band-tailed 58
Pale-vented 58
Red-billed 59
Short-billed 59
Piha, Rufous 165
Plumeleteer, Bronze-tailed 44
Potoo, Common 19
Great 20
Puffbird, White-whiskered 107

Quail-dove, Olive-backed 61
Ruddy 61
Quetzal, Resplendent 95

Rail, Grey-cowled Wood 64

Sabrewing, Violet 44
Saltator, Buff-throated 212
Sandpiper, Spotted 69
Sapphire, Blue-throated 55
Scythebill, Brown-billed 127
Seedeater, Variable 216
Sicklebill, White-tipped 22
Silky-flycatcher, Long-tailed 169
Snowcap 45
Solitaire, Black-faced 178
Spadebill, White-throated 147
Sparrow, Black-striped 188
Olive 188
Orange-billed 189
Rufous-collared 190
Stripe-headed 188
Spinetail, Slaty 131
Spoonbill, Roseate 74
Starthroat, Long-billed 35
Plain-capped 35
Stilt, Black-necked 68
Stork, Wood 71
Sunbittern 70
Sungrebe 63
Swallow, Blue-and-white 171
Mangrove 171
Northern Rough-winged 172
Southern Rough-winged 172
Swift, White-collared 20

Tanager, Black-cheeked Ant 207
Blue-grey 218
Cherrie's 215
Golden-hooded 219
Grey-headed 213
Palm 219
Scarlet-rumped 215
Silver-throated 220
Spangle-cheeked 220
Speckled 218
Summer 206
Tooth-billed 207
White-lined 214
Teal, Blue-winged 12
Thick-knee, Double-striped 66
Thorntail, Green 30
Thrush, Clay-coloured 183
Mountain 181
Pale-vented 182
Sooty 181
Swainson's 180
White-throated 182
Wood 178
Tinamou, Great 10
Highland 11
Little 11
Tityra, Masked 164
Tody-flycatcher, Black-headed 146
Common 146
Toucan, Keel-billed 112
Yellow-throated 111
Toucanet, Emerald 109
Treehunter, Streak-breasted 129
Treerunner, Ruddy 131
Trogon, Baird's 97
Black-headed 97
Black-throated 100
Collared 98
Gartered 99
Orange-bellied 98
Slaty-tailed 96
Turnstone, Ruddy 69
Tyrannulet, Mistletoe 144
Paltry 144

Umbrellabird, Bare-necked 157

Violetear, Brown 27
Mexican 28
Vireo, Mangrove 167
Red-eyed 167
Vulture, Black 82
King 81
Turkey 81

Warbler, American Yellow 200
Black-cheeked 203
Black-eared 204
Black-throated Green 202
Blackburnian 201
Blue-winged 198
Buff-rumped 202
Chestnut-sided 201
Golden-crowned 203
Golden-winged 198
Mangrove 200
Tennessee 199
Wilson's 204
Waterthrush, Northern 198
Whimbrel, Hudsonian 68
Whistling-duck, Black-bellied 12
Whitestart, Collared 205
Slate-throated 205
Woodcreeper, Barred 125
Black-striped 126
Cocoa 125
Northern Barred 125
Olivaceous 123
Plain-brown 123
Spotted 126
Streak-headed 127
Strong-billed 124
Wedge-billed 124
Woodnymph, Crowned 45
Woodpecker, Acorn 112
Black-cheeked 113
Chestnut-coloured 115
Golden-naped 113
Golden-olive 115
Hairy 114
Hoffmann's 113
Lineated 116
Pale-billed 116
Red-crowned 113
Rufous-winged 114
Woodstar, Magenta-throated 39
Wren, Band-backed 172
Bay 174
Grey-breasted Wood 176
House 175
Ochraceous 175
Plain 175
Riverside 173
Rufous-backed 173
Stripe-breasted 174
Timberline 175
White-breasted Wood 176

Xenops, Plain 128
Streaked 128

Yellowthroat, Grey-crowned 199